COMPUTERS AND DNA

COMPUTERS AND DNA

THE PROCEEDINGS OF THE
INTERFACE BETWEEN COMPUTATION
SCIENCE AND NUCLEIC ACID
SEQUENCING WORKSHOP,
HELD DECEMBER 12 TO 16, 1988
IN SANTA FE, NEW MEXICO

Edited by

George I. Bell
Los Alamos National Laboratory

Thomas G. Marr
Los Alamos National Laboratory

Proceedings Volume VII

SANTE FE INSTITUTE
STUDIES IN THE SCIENCES OF COMPLEXITY

Routledge
Taylor & Francis Group

NEW YORK AND LONDON

Director of Publications, Santa Fe Institute: *Ronda K. Butler-Villa*
Technical Assistant, Santa Fe Institute: *Della Ulibarri*

First published in 1990 by Westview Press

Published 2018 by Routledge
605 Third Avenue, New York, NY 10017
4 Park Square, Milton Park, Abingdon, Oxon OX14

Routledge is an imprint of the Taylor & Francis Group, an informa business

Library of Congress Cataloging-in-Publication Data

Interface between Computation Science and Nucleic Acid Sequencing
Workshop (1988 : Santa Fe, N.M.)
    Computers and DNA : the proceedings of the Interface between
Computation Science and Nucleic Acid Sequencing Workshop, held
December 12 to 16, 1988 in Santa Fe, New Mexico / editors, George
I. Bel, Thomas G. Marr.
    p.   cm.—(Santa Fe Institute studies in the sciences of
complexity ; proceedings v. 7)
    "The Advanced book program."
    Includes bibliographical references.
    ISBN 0-201-51505-9.—ISBN 0-201-51561-X (pbk.)
    1. Nucleotide sequence—Data processing—Congresses.   I. Bell,
George I., 1926–   . II. Marr, Thomas G.   III. Title.   IV. Series:
Santa Fe Institute studies in the sciences of complexity ; v. 7.
QP624.I53   1990   574.87'328'0285—dc20   90-27171

ISBN 13: 978-0-201-51561-9 (pbk)

This volume was typeset using T$_E$Xtures on a Macintosh II computer.

About the Santa Fe Institute

The *Santa Fe Institute* (SFI) is a multidisciplinary graduate research and teaching institution formed to nurture research on complex systems and their simpler elements. A private, independent institution, SFI was founded in 1984. Its primary concern is to focus the tools of traditional scientific disciplines and emerging new computer resources on the problems and opportunities that are involved in the multidisciplinary study of complex systems—those fundamental processes that shape almost every aspect of human life. Understanding complex systems is critical to realizing the full potential of science, and may be expected to yield enormous intellectual and practical benefits.

All titles from the *Santa Fe Institute Studies in the Sciences of Complexity* series will carry this imprint which is based on a Mimbres pottery design (circa A.D. 950–1150), drawn by Betsy Jones.

Santa Fe Institute Studies in the Sciences of Complexity

PROCEEDINGS VOLUMES

Volume	Editor	Title
I	David Pines	Emerging Syntheses in Science, 1987
II	Alan S. Perelson	Theoretical Immunology, Part One, 1988
III	Alan S. Perelson	Theoretical Immunology, Part Two, 1988
IV	Gary D. Doolen et al.	Lattice Gas Methods of Partial Differential Equations
V	Philip W. Anderson et al.	The Economy as an Evolving Complex System
VI	Christopher G. Langton	Artificial Life: Proceedings of an Interdisciplinary Workshop on the Synthesis and Simulation of Living Systems
VII	George I. Bell & Thomas G. Marr	Computers and DNA

LECTURES VOLUMES

Volume	Editor	Title
I	Daniel L. Stein	Lectures in the Sciences of Complexity

Contributors to This Volume

C. Barnes, *Los Alamos National Laboratory*

George I. Bell, *Los Alamos National Laboratory*

Craig J. Benham, *Mount Sinai School of Medicine*

A. L. Brugge, *AMOCO Technology Company*

Christian Burks, *Los Alamos National Laboratory*

Iva H. Cohen, *Yale–Howard Hughes Medical Institute Human Gene Mapping Library*

J. Collins, *Edinburgh University, Scotland*

Daniel B. Davison, *Los Alamos National Laboratory*

Charles DeLisi, *Mount Sinai School of Medicine*

R. Doolittle, *University of California, San Diego*

R. Farber, *Los Alamos National Laboratory*

Louis Gordon, *University of Southern California*

Robert Jones, *Thinking Machines Corporation and Whitehead Institute for Biomedical Research*

Kenneth K. Kidd, *Yale–Howard Hughes Medical Institute Human Gene Mapping Library*

Kimberle Koile, *Unisys Paoli Research Center*

Andrzej K. Konopka, *National Institutes of Health, National Cancer Institute*

David Kristofferson, *BIONET, IntelliGenetics*

Eric Lander, *Whitehead Institute for Biomedical Research and Harvard University*

Allen Lapedes, *Los Alamos National Laboratory*

M. N. Liebman, *AMOCO Technology Company*

John Joseph Loehr, *Los Alamos National Laboratory*

Jill P. Mesirov, *Thinking Machines Corporation*

Sanzo Miyazawa, *National Institute of Genetics, Japan*

K. Nakata, *National Cancer Institute*

G. Christian Overton, *Unisys Paoli Research Center*

John Owens, *National Institutes of Health, National Cancer Institute*

Jon A. Pastor, *Unisys Paoli Research Center*

S. Reddaway, *Active Memory Technology Ltd.*

Temple F. Smith, *Dana-Farber Cancer Institute*

Karl M. Sirotkin, *Los Alamos National Laboratory*

J. Claiborne Stephens, *Yale–Howard Hughes Medical Institute Human Gene Mapping Library*

Washington Taylor IV, *Thinking Machines Corporation and University of California, Berkeley*

David C. Torney, *Los Alamos National Laboratory*

Michael S. Waterman, *University of Southern California*

Xiru Zhang, *Brandeis University*

George I. Bell
Los Alamos National Laboratory, Los Alamos, NM 87545

Preface

The fields of molecular biology and genetics are faced with the accumulation of quantitative information at an ever increasing rate such that the unaided human mind cannot begin to assimilate or analyze its significance. The chief source of this information is DNA sequencing and, thereby, the associated sequences of amino acids in proteins; but genetics, macromolecular structure, and other data sets are also large. Because of the great importance of these data, molecular biologists have turned to computational scientists for help in organization of the data into accessible and current databases and for software and algorithms for its analysis. Programs to map and sequence complex genomes, including bacteria, yeast, and human, are underway; and advances in computational science will be required to keep pace with the molecular biology.

In order to discuss current approaches and needs as well as to introduce computational scientists to the field, the Santa Fe Institute organized a workshop on "The Interface between Computational Science and DNA Sequencing," held in Santa Fe, New Mexico, December 12–16, 1988. Approximately one hundred molecular biologists, computer scientists, mathematicians, and other scientists in diverse fields attended. The papers in this volume were presented at that meeting and are intended to present both an introduction to the field and a discussion of research on some of the current problems.

The workshop was supported by the Department of Energy, Office of Health and Environmental Research, and the Los Alamos National Laboratory (LANL).

Much of the organization and running of the workshop was done by Tom Marr (LANL), ably assisted by Andi Sutherland and Ginger Richardson (SFI) and Annette Martinez (LANL). Manuscripts were typed for publication by Ronda K. Butler-Villa.

George I. Bell
Los Alamos, NM
June 1989

Contents

Introductory

George I. Bell
Center for Human Genome Studies, Los Alamos National Laboratory, Los Alamos, NM 87545

The Human Genome: An Introduction

DNA

A human genome consists of the deoxyribonucleic acid (DNA) content of a human cell, and it contains all the information that is required for the development of a human being from a fertilized egg, a single cell. DNA is a linear polymer, composed of four nucleotides or bases, A (adenine), C (cytosine), G (guanine), and T (thymine), that are linked together by a sugar-phosphate backbone (see Figure 1). The order of the bases is the information in the DNA, determining its function. Thus a DNA molecule is a message written in a four-letter alphabet.

DNA is generally found in the form of a double helix, two intertwined DNA molecules in which the order of the bases on one of the molecules is complementary to that on the other. Thus, wherever there is an A on one strand, there will be a T on the other; and similarly, G will be found paired with C. This complementarity is determined by the pattern of hydrogen bonding between A and T or G and C, as shown in Figure 2. Prior to cell division, the DNA in a cell must be precisely duplicated; this is done by unwinding the double helix and synthesizing two new complementary strands.

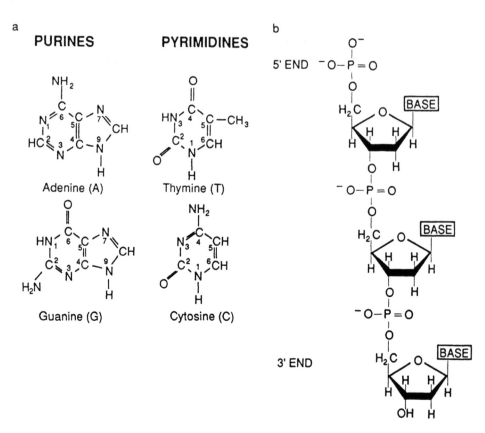

a

PURINES   PYRIMIDINES

Adenine (A)

Thymine (T)

Guanine (G)

Cytosine (C)

b

5' END

3' END

FIGURE 1   a. The four bases, A, T, G, and C, are shown. b. They are linked together by a sugar-phosphate backbone to form the linear DNA polymer.

Complementarity of interacting DNA strands is exploited by molecular biologists seeking a particular short DNA sequence. The process is to prepare a short strand of radioactive complementary DNA, which will, under suitable conditions, bind only to the single DNA strand of interest. Binding is then visualized by exposing the radioactivity with a photographic film. Alternatively, the binding may be visualized by a fluorescent or chemiluminescent signal.

A human cell contains approximately 3×10^9 base pairs from each parent. If depicted on a printed page, the message would extend for about a million pages. The DNA is organized into 23 pairs of chromosomes, 22 pairs of autosomes—with one member of each pair from each parent—and 2 sex chromosomes—2 X chromosomes for a female and an X and Y for a male. A chromosome may be regarded as a single molecule (actually a double helix) of DNA, although the DNA in a cell is generally associated with a variety of protein molecules that modulate its structure and function.

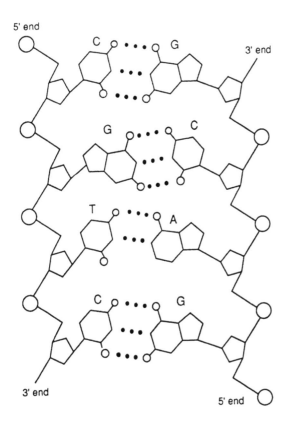

FIGURE 2   In the Watson-Crick double helix, two anti-parallel strands of DNA are present such that there are hydrogen bonds between C on one strand and G on the other, and similarly between A and T. The hydrogen bonds are depicted by dark dots.

The length of the DNA in a human cell is about one meter from each parent. Yet it is packed into the nucleus of a cell having a diameter of about $5\mu m$. The compaction of DNA into such a small volume, while preserving its organization such that in a cell division each daughter cell receives precisely half of the DNA, indicates a high level of organization in the DNA, probably maintained by its interaction with certain proteins that form a scaffold.

Approximately one tenth of one percent of the DNA in a human cell has been sequenced.

GENES AND PROTEINS

A gene is a unit of heredity, and most genes are now known to be segments of DNA that specify particular protein molecules. There are estimated to be of the order of 10^5 genes in a human cell of which about one percent are characterized. Defects in single genes have been associated with more than four thousand diseases, including cystic fibrosis, Huntington's disease, and hemophelia. In addition, many common diseases, such as coronary disease, hypertension, certain kinds of mental disease, and cancer, have strong genetic components; disease susceptibility is probably influenced by several genes and environmental factors as well. A major objective of the human genome program is to locate and characterize genes responsible for genetic disease, thereby accelerating understanding and possible treatment of the diseases.

Proteins are also linear polymers composed from 20 different kinds of monomers, called amino acids. The *genetic code* specifies how the order of the bases in DNA determines the order of the amino acids in the protein. In essence, the sequence of these bases in DNA determines which of the amino acids will appear in the protein; complex molecular machinery involving many kinds of protein and RNA (ribonucleic acid) carry out this process. The genetic code is shown in Figure 3.

The various amino acids differ as to size, charge, and solubility; hence, although a protein is a linear polymer, in water it will fold up into a complex three-dimensional structure that will determine its function, perhaps as an enzyme, a structural member, or a molecule that binds to specific DNA sequences.

Only a small fraction of the human genome codes for proteins. A quantitative estimate may be made as follows. Suppose that the average protein has about 300 amino acids and is thus coded by about 1000 bases. On 10^5 genes, the total coding DNA would thus be about 10^8 bases or three percent of the total genome.

First Position	Second Position				Third Position
	U	C	A	G	
U	Phe	Ser	Tyr	Cys	U
	Phe	Ser	Tyr	Cys	C
	Leu	Ser	STOP	STOP	A
	Leu	Ser	STOP	Trp	G
C	Leu	Pro	His	Arg	U
	Leu	Pro	His	Arg	C
	Leu	Pro	GLuN	Arg	A
	Leu	Pro	GLuN	Arg	G
A	Ileu	Thr	AspN	Ser	U
	Ileu	Thr	AspN	Ser	C
	Ileu	Thr	Lys	Arg	A
	Met	Thr	Lys	Arg	G
G	Val	Ala	Asp	Gly	U
	Val	Ala	Asp	Gly	C
	Val	Ala	Glu	Gly	A
	Val	Ala	Glu	Gly	G

FIGURE 3 The Genetic Code.

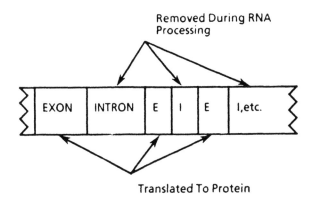

FIGURE 4 Structure of
typical eukaryote gene.

What is the function of the rest of the genome? A considerable fraction is found in segments of DNA that interrupt the coding sequences of genes; they are known as intervening sequences, or introns, and their discovery astonished molecular biologists. Thus, in human cells and those of all multicellular creatures, the genomic DNA sequence coding for a gene is not colinear with the amino acid sequence of the resulting protein, but is interrupted by introns (see Figure 4). Some genes contain only a few introns, others more than 50. Prior to translation of the DNA message into a protein, an RNA (ribonucleic acid) copy of the DNA is made, and introns are removed from the RNA by a complex splicing mechanism such that the resulting *messenger* RNA copy *is* colinear with the translated proteins.

The function of introns is controversial. Expressed gene segments, known as exons, are often thought to represent functional domains or motifs in the protein, and this appears to be true for at least some exons. Genes contain roughly ten times more DNA in introns than in coding sequences. Thus we may estimate that introns amount to ~ 30 percent of the human genome.

About another 30 percent codes for repetitive sequences, including some that are associated with the ends of chromosomes (telomeres) or with those regions near the center (centromers) that are used to pull chromosomes apart in cell division. Other repetitive sequences are interspersed throughout the genome, and some of these appear to represent parasitic DNA, that is, segments that can become replicated and transposed throughout the genome. Some of the repetitive sequences may function to maintain chromosome structures or as origins of DNA replication.

MAPS

Major elements of the Human Genome Program concern the determination of maps of the 24 kinds of human chromosomes at various degrees of resolution. Since the chromosomal DNA molecules are linear polymers, a map of a chromosome will be an ordered linear array of genes, markers, or nucleotides. The various kinds of maps differ as to the resolution of the representation and as to the experimental methods that are used in the mapping process (see Figure 5).

A gross physical map or cytogenetic map is obtained by locating various markers on a chromosome using techniques including *in situ* hybridization and the study of cells that retain only portions of the chromosome. Chromosomes can be observed in a microscope during mitosis, or cell division, and suitably tagged segments of DNA can be visualized as they bind to complementary sequences on the chromosome. In this process of *in situ* hybridization, the resolution is typically of the order of 10^7 base pairs, or ten percent of a typical chromosome's length. In another approach, use is made of cell populations that by design contain only a portion of the chromosome under study. By examining whether or not two markers are on the same fragment, it may be possible to determine the spacing between markers to $\sim 10^6$ base pairs or better. Moreover, recent progress in *in situ* hybridization has raised the possibility of resolving markers spaced by $\sim 10^6$ base pairs in that approach as well.

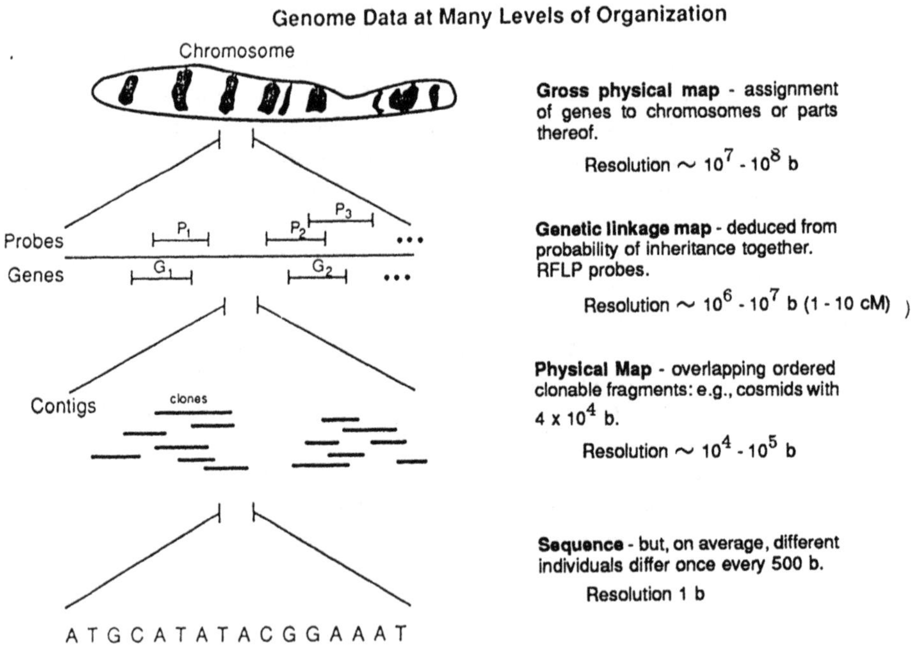
Genome Data at Many Levels of Organization
Chromosome
Gross physical map - assignment of genes to chromosomes or parts thereof.
Resolution ~ 10^7 - 10^8 b
Probes
P1
P2
P3
Genes
G1
G2
Genetic linkage map - deduced from probability of inheritance together. RFLP probes.
Resolution ~ 10^6 - 10^7 b (1 - 10 cM)
Contigs
clones
Physical Map - overlapping ordered clonable fragments: e.g., cosmids with 4 x 10^4 b.
Resolution ~ 10^4 - 10^5 b
Sequence - but, on average, different individuals differ once every 500 b.
Resolution 1 b
A T G C A T A T A C G G A A A T

FIGURE 5 Various kinds of maps will be important in understanding the human genome.

GENETIC LINKAGE MAPS

A different kind of map is obtained by studying the coinheritance of genes or other markers in human families. By this means it was first established that certain genes lie on the human X chromosome—for example, those involved in hemophelia and Duchene's muscular dystrophy. These conclusions were established by noting that such diseases occur in males, but the pattern of inheritance shows that females are the carriers. When two or more genes or markers are located on the same chromosome, a distance between them can be estimated from the probability that they are not inherited together. This approach is made possible by the phenomenon of crossing over during meiosis. Meiosis refers to the cell division that is involved in the formation of germ cells, sperm, or egg. In this cell division, the number of chromosomes passed on to each daughter cell must be halved so that the daughter cell receives 22 autosomes (instead of 22 pairs) plus one sex chromosome (instead of 2). However, during the preparation for meiosis, homologous chromosomes pair up and frequently exchange genetic material via crossing over (Figure 6). The probability that two linked (on the same chromosome) markers become separated is called the genetic distance between them and is measured in centiMorgans (cM). One cM corresponds to a probability of one percent recombination and is roughly 10^6 base pairs. The length of a human chromosome is of the order of 100 cM; thus, there is a substantial probability of a crossover event involving each human chromosome.

Markers that appear in a genetic linkage map can be disease genes or any kind of marker that is detectable and polymorphic (has many forms in different individuals in the population). In particular, it is possible to cut up human DNA with various enzymes (restriction enzymes) that cut only at specific sequences (typically of 4 or 6 base pairs), sort the fragments according to size, and visualize specific fragments

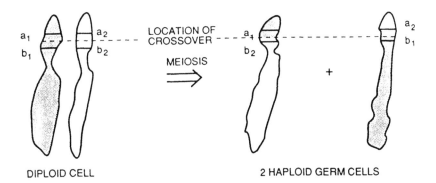

FIGURE 6 During crossing over, two markers, a and b, that were originally on the same chromosome may become separated if the crossover occurs between them.

by their binding to probe nucleic acid sequences. Individuals may differ in their distribution of fragment lengths either because a restriction site present in one individual may be absent in another or because the distance between sites may differ in various individuals. In this way, several hundred anonymous markers have been mapped on human chromosomes. Overall, the distance between markers is about 5 cM, or 5×10^6 base pairs. A near term objective of the Human Genome Program is to refine the genetic linkage map to the 1 cM level.

Inasmuch as the primary data for composing a genetic linkage map are *probabilities* of coinheritance of linked markers, as determined from the limited number of families studied, there are numerous statistical problems in generating the map. It is believed that the probability of recombination is not uniformly distributed along a chromosome but is concentrated in various hot spots or regions. Thus the transformation from genetic linkage distances to numbers of base pairs is very approximate.

Genetic mapping methods have indicated that individuals are highly polymorphic, differing at perhaps one base in 500. It is believed that most of those differences do not lie in genes but in introns and other regions of the genome, sometimes called junk DNA.

PHYSICAL MAPPING

A physical map of a chromosome is a linear array of markers along the chromosome in which the distance between markers is proportional to the number of bases between them. The sequence itself is, of course, the ultimate physical map, with resolution at the single base level. However, coarser physical maps, namely ordered sets of clonable fragments, will play key roles in the Human Genome Program. A near term objective is to obtain such a physical map of each human chromosome. The number of elements in a map and its resolution depends on the length of the cloned DNA fragments. Several cloning systems are in use including λ phage (length $\sim$ 10 kb), cosmid ($\sim$ 40 kb), and the yeast artificial chromosome (or YAC), currently yielding lengths $\sim$ 300 kb. Thus the number of clones required to cover a typical chromosome of length $\sim$ 100 Mb would be roughly λ (10^4 clones), cosmid (2500), and YAC (333).

In constructing a physical map of a chromosome, the current procedure is to start with the DNA from a single human chromosome, obtained by sorting the chromosomes in a flow cytometer, using a hybrid cell line with but a single human chromosome or some other method. This DNA is then cut up into fragments of appropriate size, e.g., using restriction enzymes, which are incorporated into suitable vectors and cloned. To obtain a coverage of most of the chromosome DNA, one needs enough clones to span the chromosome several times.

It is next necessary to determine which clones overlap and to order them into a map. This can be done by partially characterizing each clone since, if two clones have enough properties in common, they may be concluded to overlap, that is, contain a section of DNA in common. Clone characteristics that may be useful include hybridization to defined DNA segments such as short oligonucleotides or repetitive sequences and/or the distribution of fragment lengths when the clone is cleaved by restriction fragments. Alternatively, the hybridization of clones to each other may be used to indicate sharing of DNA.

ORGANIZATION OF THE DATA

Extensive mapping data will be generated in the Human Genome Project, especially that concerning the genetic linkage maps, physical maps, and the sequences. In a 1 cM genetic linkage map there would be $\succeq 10^4$ markers; in a cosmid scale physical map, more than 10^5 clones; and in a sequence database, more than 10^{10} bases if representation of some human polymorphism is included. All of this information must be organized into a friendly computer database or, if not a single database, then a set of databases so closely linked or cross-referenced as to appear unified to the user.

ANALYSIS OF THE DATA

The *sequence* of a human genome should be regarded as only a beginning in *understanding* the genome. In particular, scientists concerned with human genetic disease, or with normal human development and function, will wish to identify all the genes, to understand their regulation and the structure and function of their translation products, the proteins. A paradigm is furnished by the task of deducing the function of a protein from the genomic DNA sequence of the underlying gene.

REFERENCES

Of special relevance to the Human Genome Program are:

1. Committee on Mapping and Sequencing the Human Genome. *Mapping and Sequencing the Human Genome.* Washington, D.C.: Nat. Acad. Press, 1988.
2. U.S. Congress, Office of Technology Assessment. "Mapping Our Genes—The Genome Projects, How Big, How Fast?" OTA-BA-373. Washington, D.C.: U.S. Govt. Printing Office, 1988.

More general references are:

3. Alberts, B., D. Bray, J. Lewis, M. Raff, K. Roberts, and J. D. Watson. *Molecular Biology of the Cell.* New York: Garland Publishing, 1983.
4. Darnell, J. J., H. Lodish, and D. Baltimore. *Molecular Cell Biology.* New York: W. H. Freeman, 1986.
5. Watson, J. D., L. Tooze, and D. T. Kurtz. *Recombinant DNA, A Short Course.* New York: W. H. Freeman, 1983.

Charles DeLisi
Department of Biomathematical Sciences, Mount Sinai School of Medicine, New York, NY 10028

Computation and the Human Genome Project: An Historical Perspective

INTRODUCTION

One of the striking characteristics of the human genome project, and one of its more appealing aspects as a social enterprise, is the extraordinarily diverse backgrounds of the individuals that have been drawn to it. I am occasionally reminded of *Cat's Cradle*, the novel by Kurt Vonnegut in which the book of Bokonon is described. A central Bokononian concept is the *karass*—a group of otherwise unrelated individuals united by a common object—not unlike scientists in search of the genome. A wampeter is the object that unites the karass. What is interesting is that Vonnegut cites the Holy Grail as an example of a wampeter. Although I convey these observations with some frivolity, the connection with the famous phrase used to describe our quest for the genome is more than coincidental and far from frivolous: there is a fundamental symbolism in this project which Walter Gilbert has succinctly captured in his allusion to the grand quest of Perceval Le Gallois. But let me avoid over-intellectualizing and move more directly to the topic of this talk, which concerns the scientific history of the computational components of the project.

Although I am limiting my comments to computation, I am still not going to be able to be complete, nor will I make pretense at objectivity. I will focus on only a single and perhaps obvious theme and draw only on those examples needed to provide minimal illustration. The theme is that major advances in experimental technologies during the past quarter century have opened new research areas with substantial computational components. Furthermore, we are about to witness what will be by far the most explosive increase in technological developments and data generation that has ever occurred in the biological sciences. A corollary, therefore, is that we can expect an increasingly prominent role for theory and advanced computation.

ANCIENT HISTORY

Although this workshop is on DNA sequencing, the initial stimulus to mathematical and computational analysis was the growth of protein sequences in the 1960's. At least two observations are pertinent. (1) Analysis of sequences alone can provide important biological insights. (2) A point is quickly reached at which the number of sequences becomes too large to either manage *or analyze* without recourse to computers. Both are obvious, and yet neither seems to have been sufficiently internalized to make the impact on science policy needed to prepare us for optimal progress in the future.

The work of Wu and Kabat[18] illustrates what we can learn from sequence analysis. When immunoglobulin sequences are aligned, and the number of residue substitutions is plotted as function of position, most of the changes are localized to three *hypervariable regions.* Since the immune system must be able to recognize any of the virtually unlimited number of antigens with which it can come in contact, antibody binding sites are expected to be highly heterogeneous. The results therefore suggest that the hypervariable regions contribute to the combining site. In addition, with only one site per antibody, the molecule must fold so that the regions, though separated along the chain, are spatially proximate. Subsequent x-ray diffraction analysis confirmed the conclusions of Wu and Kabat.

More generally, a number of protein databases were developed in the 1960's and 1970's to help manage and analyze the increasingly large amount of protein sequence information.[2] But unlike the analysis of immunoglobulins, whose variable region domains are all functionally similar, comparison of protein sequences that might be less obviously related, especially comparisons made in developing evolutionary trees, requires a precise definition of similarity, a requirement that stimulated pioneering mathematical work on distance metrics by Ulam and Beyer at Los Alamos.[14] Also unlike comparison of immunoglobulins, whose number is relatively limited, a search and alignment procedure in which an entire database is used, requires computerized alignment algorithms.

One of the earliest and most influential sequence alignment algorithms was developed by Needleman and Wunch.[11] It is based on the principle of dynamic programming, and it provided the basis for many subsequent developments, including a method for finding the tertiary structure of peptides, which I will describe later. Moreover, under the assumptions of their model, the method rigorously finds the best alignment of two sequences.

A problem related to finding the optimal alignment of two proteins is finding the secondary structure of an RNA molecule. In the 1960's, RNA was considerably easier to sequence than DNA, and a large number of tRNA sequences had become available, as well as sequences of a number of other RNA molecules, some several hundred bases long. These single-stranded molecules can assume various secondary structures by looping back on themselves in a hairpin configuration, by forming loops internal to a helix, or by bulging on one side of a helix as a result of misaligned bases.[15] The various loops have size-dependent free energies, which at the time had to be obtained theoretically using statistical mechanical methods.[3] Unlike the problem of merely counting aligned residues to weight a configuration, the problem of RNA structure determination required taking account of distance-dependent free energies. This complicates the algorithm considerably. Shortly after I arrived at Los Alamos in the early 1970's, I met Michael Waterman. The problem caught his attention and he went on to pioneer the development of secondary structure algorithms[17] and to solve a number of other important problems in computational molecular biology.

Interest in theoretical biology at Los Alamos grew rapidly throughout the 1970's. George Bell formed the Theoretical Biology and Biophysics group; Minoru Kanehisa and Walter Goad developed methods for linking, searching, and manipulating databases; Temple Smith and Mike Waterman developed and applied new methods for DNA and RNA sequence and structure analysis; some of the more difficult combinatorial problems were holding the interest of outstanding probability theorists like Paul Stein and Gian Carlo Rota. It was a time of great excitement, and in a sense it primed theorists for the changes that were about to occur—changes driven by technological breakthroughs in experimental methodologies made by Gilbert and Maxim in the United States, and by Sanger in England.

The development of high-speed DNA sequencing methods increased the flow of data several hundred-fold, and the management problem quickly became severe enough to attract the attention of the National Institute of General Medical Sciences, the primary funding organization for basic research in molecular biology. Several meetings were held to discuss the development of a computerized database, the ultimate result being the award of a contract to Bolt Berenek and Newman with a subcontract to Los Alamos. The Theoretical Biology and Biophysics group became the site of GenBank.

CURRENT EVENTS

The data explosion continued through the early 1980's, and by 1985 GenBank had a total of 6 million bases, approximately 1 million of which represented human DNA.

The thought of sequencing the entire human genome inevitably occurred to those given to mathematical extrapolation. The history of what is now called the human genome project, the ultimate goal of which is to obtain the complete sequence of human DNA, is well known, having been told many times,[6] and I will not recount it here. Of greater interest is its implications for computational biology.

In the early 1980's, well before anyone in this country seriously considered sequencing the entire human genome, simple projections of data accumulation rates indicated a need for more rapid methods for determining the function of DNA sequences and the cellular location, function, and structure of gene products.[5] The currently anticipated increase in the amount of data will have two effects on these problems: they will be greatly exacerbated and, equally importantly, the likelihood of developing reliable computer methods for their solution will increase, since the methods must be developed using known structure function relations.

The problem of computer-assisted functional identification is not new: it is the central objective of homology search algorithms.[10] Complete database searches using these algorithms will, however, become increasingly less effective as the sequence explosion evolves. Alternative approaches involve the identification of simple motifs of structure and function. Some of these methods will be reviewed elsewhere in this volume by Doolittle.[8] I only note here that several different approaches are being pursued. These involve functional classification in terms of local sequence properties[1,7,9]; in terms of disulfide bond topologies[8]; and in terms of topological arrangements accessible to proteins rich in beta structure.[13]

Methods that rely only on local sequence motifs, or even general sequence properties, are not likely to be fully reliable. A simple example which illustrates when these methods are expected to succeed and fail is the prediction of antigenicity. Determinants that bind to immunoglobulin receptors on antibody-producing B cells are formed by the spatial juxtaposition of different portions of a protein, i.e., by spatial proximity of residues that are separated along the sequence. Since antibodies generally recognize geometric characteristics of the folded protein, identification of residues that are candidates for binding the B cell's antibody-like receptors is very difficult unless the fold is known. For example, predictions based on the idea that contact residues must be on the protein surface and are therefore expected to be relatively hydrophilic are only marginally successful.

In contrast to predicting B-cell antigenicity, predicting T-cell antigenicities is relatively reliable. The reason is that T cells recognize *local* rather than long-range sequence properties. Native protein is ingested by a B cell, macrophage or dendritic cell, enzymatically cleaved, and recycled to the surface in the form of short peptides (10–15 residues long) that are recognized in conjunction with products of the major histocompatibility gene complex (MHC). A number of observations related to this process led to the hypothesis that T-cell immunodominant peptides are amphipathic alpha helices.[4] Data accumulated over the past three years indicate that approximately 70% of peptides known to stimulate T helper cells are alpha-helical amphipathic structures.

Alpha amphipathicity is likely only one of several structural motifs assumed by immunodominant peptides. However, since a very limited number of MHC products (in a given individual) must be able to associate with a very large number of peptides, we expect that the number of peptide motifs must also be very limited. The motifs are essentially intrinsic to the peptide, but a large number of neighboring structures are also expected, these being imposed by the requirement that a given peptide be able to bind highly polymorphic gene products.

The search for antigenic motifs, or more generally the development of a predictive understanding of the requirement for recognition (e.g., predicting the antigenic effect of a residue substitution on any of the components in the ternary complex), is a central goal of molecular immunology. Because of the large numbers of peptides, MHC products, and T cell receptors, the problem must be approached theoretically; i.e., a general algorithm, if not a principle, is required that will allow an accurate determination of the antigenicity of any given peptide in any specified system. We have recently developed generalized dynamic programming algorithms[16] that permit determination of peptide structures in less than an hour on a VAX 11/780, opening the way for routine determination of the structure of such molecules. Moreover, the generic fold of MHC products is now known from the crystal structure of a class 1 product. Energy minimization methods will soon be available for obtaining neighboring structures, thus making possible the determination of the precise structure of any specified class 1 or class 2 product. The longer term goal, which should be achieved in the next 3–5 years, is accurate calculation of the structure of the complex.

This example is perhaps a verbose way of stating the obvious: sequence alone *can* provide clues to generic function or activity, but only some of the time. A general approach requires specification of tertiary structure.

THE NEXT DECADE

By the turn of the century or shortly afterward, we can expect to see a complete human genome sequence, probably also a murine sequence, and several high-resolution human gene maps. Of equal importance, the protein-folding problem, which is crucially significant for a predictive understanding of activity, will be essentially solved and the relation between structure and generic function will have been largely unraveled, the latter permitting real-time conversion of sequence data to scientific knowledge. The emergence of vast quantities of fundamentally important data and rapid analytical capabilities, will provide major surprises of fundamental and practical importance.

These advances, of course, presuppose adequate funding, but I feel fully confident that such funding will be forthcoming. I have often stated that the human genome project has a life of its own. The reason lies in the project's symbolic significance as a fundamental human drive. It symbolizes a story that has been told many times before, the story of the grand quest, indelibly captured over eight centuries ago in Chretien de Troyes' famous poem which sings of the search for the chalice of Christ. But it symbolizes much more. It expresses in a very powerful fashion a collective drive to understand the essence of life; to understand its evolution; to understand ourselves as individuals and our biological linkage to the awesome diversity of life that surrounds us—and to do so at a depth that the Delphic Oracle could scarcely have begun to imagine.

REFERENCES

1. Blundell, T. L., B. L. Sibanda, M. J. E. Sternberg, and J. M. Thornton. "Knowledge-Based Prediction of Protein Structures and the Design of Novel Molecules." *Nature* **326** (1987):347.
2. Dayhoff, M. O. *Atlas of Protein Sequence and Structure.* Washington, DC: Georgetown University Press, 1978.
3. DeLisi, C., and D. M. Crothers. "Prediction of RNA Secondary Structure." *Proc. Nat. Acad. Sci. USA* **68** (1971):2682.
4. DeLisi, C., and J. Berzofsky. "T-Cell Antigenic Sites Tend to be Amphipathic Structures." *Proc. Nat. Acad. Sci. USA* **82** (1984):140.
5. DeLisi, C., P. Klein, and M. Kanehisa. "Some Comments on Protein Taxonomy: Procedures for Functional and Structural Classification." In *Mol. Basis of Cancer: Macromolecular Structure, Carcinogens, and Oncogenes.* New York: Alan R. Liss, 1985, 341–441.
6. DeLisi, C. "The Human Genome Project." *The Amer. Scientist* **76** (1988):488.
7. DeLisi, C. "Computers.in Molecular Biology: Current Applications and Emerging Trends." *Science* **240** (1988):47.
8. Doolittle, R. F. This volume.
9. Kelin, P., M. Kanehisa, and C. DeLisi. "Prediction of Protein Function by Discriminant Analysis." *Math Biosci.* **81** (1986):177.
10. Lipman, D. J., and W. R. Pearson. "Rapid and Sensitive Protein Similarity Searches." *Science* **227** (1985):1435.
11. Needleman, S. G., and C. D. Wunsch. "A General Method Applicable to the Search for Similarities in the Amino Acid Sequence of Two Proteins." *J. Mol. Biol.* **48** (1970):443.
12. Posfai, J., A. S. Bhagwat, G. Posfai, and R. J. Roberts. "Predictive Motifs Derived from Cytocine Methytransferases." *Nucl. Acids Res.*, submitted.
13. Richardson, J. "β-Sheet Topology and the Relatedness of Proteins." *Nature* **268** (1977):495.
14. Smith, T. F., W. A. Beyer, and M. Waterman. "Some Biological Sequence Metrics." *Adv. in Math.* **20** (1976):367.
15. Tinoco, I., O. Uhlenbech, and M. D. Levine. "Estimation of Secondary Structure in Ribonucleic Acids." *Nature* **230** (1971):362.

16. Vajda, S., and C. DeLisi. "Predicting Low-Energy Conformations of Short Polypeptides by Discrete Dynamic Programming." Submitted.
17. Waterman, M. "Secondary Structure of Single Stranded Nucleic Acids." *Adv. in Math. Supplementary Studies* **1** (1978):167.
18. Wu, T. T., and E. Kabat. "An Analysis of Sequences of the Variable Regions of Bence Jones Proteins and Myeloma Light Chains and Their Implications for Antibody Complementarity." *J. Exp. Med.* **132** (1970):211.

Russell F. Doolittle
Department of Chemistry, Center for Molecular Genetics, M-034, University of California, San Diego, La Jolla, CA 92093

What We Have Learned and Will Learn from Sequence Databases

INTRODUCTION

The usefulness of sequence databases for the biological sciences needs no defense. Indeed, their value has been apparent from the first efforts of Eck and Dayhoff[15] a generation ago. As a simple resource, sequence databases are drawn upon by scientists from a broad range of disciplines for a wide variety of needs. The crystallographer needs protein sequences in order to fit individual amino acids to electron density maps, the gene expressionist wants to know about DNA flanking sequences, the biotechnologist wants to fashion probes and primers on known regions of DNA, and the protein chemist wants to synthesize peptides that can simulate parts of a protein. All these investigators, and many others, use the sequence database as a standing resource. They look up specific sequences by title or accession number and retrieve the sequence and any available information about it.

But there is an even greater, more intrinsic, value of the sequence database that transcends its function as a simple knowledge-transfer resource. The databases themselves generate new knowledge. This new knowledge is providing a remarkable picture not only of how living systems have evolved, but also how they operate. For nowhere in the biological world is the Darwinian notion of "descent with modification" more apparent than in the sequences of genes and gene products. We can see now that the predominant mode of gene evolution has been "duplicate and modify." The happy result of this historical expansion is that many, perhaps most, extant proteins are still similar enough in sequence that their early histories can be reconstructed. There have already been enormous insights gained about these phenomena, and we can anticipate a much deeper and detailed understanding of events as the sequence collections grow. In this short article, I review a few of the findings that have emerged from searching and screening sequence data banks during the last decade. With that as background, I go on to make some predictions about what we can expect to find during the next decade, and in particular what can be expected from the proposed mega-sequencing associated with the Human Genome Project.

TABLE 1 Some Startling Match-Ups Found by Computer Searching[1]

Matched Proteins	Year Reported
ovalbumin/α_1-antitrypsin	1980
v-mos/v-src	1981
v-src/CAMP-kinase	1981
platelet-derived growth factor/v-sis	1983
angiotensinogen/α_1-antitrypsin	1983
yeast CDC28/v-mos	1984
v-erb B/EGF-receptor	1984
clotting Factor VIII/ceruloplasmin	1984
$\alpha-2$-macroglobulin/complement factors C3 and C4	1985
kininogens/thiol protease inhibitors	1985
LDL-receptor/EGF-precursor	1984
complement C9/LDL-receptor	1985
complement Factor B/α_2-glycoprotein	1985
glucocorticoid receptor/v-erb A	1985
β-adrenergic receptor/rhodopsin	1986
complement Factors B/H/Factor XIII b-chain	1986

[1] The listed proteins beginning with the prefix v- (as in v-mos, v-src, etc.) are all retroviral transforming factors (adapted from Doolittle[11]).

UNEXPECTED MATCH-UPS

Sequence searching during the past decade has already revealed a number of interesting match-ups (Table 1). In retrospect, some of these might have been anticipated; others were wholly unexpected. A few have proved to be strategically important to biomedical research. For example, a search of a short amino-terminal peptide sequence of the melanoma tumor antigen suggested that it might be homologous to the iron-transporting protein transferrin.[3] Direct experiments showed that the protein did indeed bind iron with the same avidity as transferrin. Eventually, a cDNA sequence was determined for the protein, and it was found to be about 35 percent identical with transferrin, leaving no doubt of common ancestry.[24] It is highly unlikely that the function of this important cancer-related protein would have been determined so quickly without the information provided by that preliminary search in 1981.

In another case, a partial sequence was published for platelet-derived growth factor, and a search of a protein sequence database revealed that it was almost identical with a portion of an oncogene known as v-*sis* (Figure 1).[9] This finding had an immediate impact on how cancer biologists viewed oncogenesis. Since that time, a number of other oncogenes have been identified or otherwise deciphered on the basis of sequence searching also, including v-*erb* B,[14] v-*mas*,[29] v-*erb* A,[27] and v-*jun*,[25] *inter alia*.

M T L T W Q G D P I P E E L Y K M L S G H S I R S F D D L Q

R L L Q G D S G K E D G A E L D L N M T R S H S G G E L E S

L A R G K R s l g s l S V a e p a m i a e c k t r T e v f E

i S r r l I d R T N A N F L V W P P C V E V Q R C S G C C N

N R N V Q C R P T Q V Q L R P V Q V R K I E I V R K K P I F

K K A T V T L E D H L A C K C E I V A A A R A V T R S P G T

S Q E Q R A K T T Q S R V T I R T V R V R R P P K G K H R K

C K H T H D K T A L K E T L G A

FIGURE 1 Amino acid sequence of the simian sarcoma virus oncogene (v-*sis*) showing where the partial amino acid sequence of Platelet Derived Growth Factor (PDGF) matched it. Residues that were identical are shown in lower case (adapted from Doolittle et al.[8]).

TABLE 2 Some Members of the "Serpin" Family

Known Protease Inhibitors	Others
α_1-Antitrypsin	Leuserpin-2
α_1-Antichymotrypsin	Ovalbumin
α_2-Antiplasmin	Angiotensinogen
Antithrombin III	Thyroxine-binding globulin
Protein C Inhibitor	Corticosteroid-binding globulin
C1-Inhibitor	Endosperm protein Z, barley

Although all of these findings were of great importance to cancer biology, unexpected matches have had an even greater impact in the more general area of molecular evolution. Even before computer searching became popular, there was evidence that some pairs of proteins with quite unrelated functions had sequences that were so similar that common ancestry had to be the case. Thus, mammalian lactalbumins were about 35% identical with avian lysozymes,[2] and haptoglobin, a scavenger protein found in blood plasma, was more than 20% identical with chymotrypsin.[1] But it was not until the systematic computer searching of sequences was undertaken in earnest that the full extent of this enslavement of one structure for use in another capacity became apparent. Typically, one unexpected relationship for a particular type of protein would be found, and then suddenly, a whole new family would materialize. For example, Hunt and Dayhoff[18] found that ovalbumin, a protein of unknown function found in avian egg whites, was homologous with two protease inhibitors, which themselves had previously been found to be related. Shortly thereafter, the hormone precursor angiotensinogen was found to belong to this family,[8] and after that a deluge (Table 2).

Similarly, a search of the sequence for the β-adrenergic receptor resulted in the astonishing finding that the protein was homologous with rhodopsin.[6] Within two years the number of known family members had grown to more than a dozen, including the α-adrenergic receptor,[20] the muscarinic acetylcholine receptor,[21] substance K receptor,[23] the serotonin receptor,[19] and a yeast mating factor.[22] Similar outpourings followed other computer-based connections, including extended families of cysteine-protease inhibitors, cell adhesion proteins, steroid-binding proteins, and a number of others.

EXON SHUFFLING

Sometimes sequence match-ups show that only portions of one protein are similar to another. Moreover, as more genomic DNA sequences became available, it was found that such similar segments were often set off by intervening sequences, and it soon became evident that certain exons—corresponding to particularly stable and compact domains—have indeed been shuffled around in the genome during evolution. Several types of these widely distributed domains are known, and they are often named after the setting in which they were first identified. Thus, there is a characteristic EGF-domain, named after epidermal growth factor, several sets of fibronectin finger domains, a β_2 glycoprotein domain, and so forth (Figure 2). Most often these shuffled segments are 40–90 amino acid residues in length. Their compactness and stability is inferred from the fact that many of them contain 2–3 disulfide bonds. So far, they are most evident in recently evolved proteins that have been assembled in their present form in the time since the divergence of animals from plants and fungi. An illustration of the wide distribution of a shuffled segment is afforded by the EGF-unit (Table 3).

Typical Occurrence	Unit Length	Disulfides	Disulfide Arrangement
Complement C9	40 ± 2	3	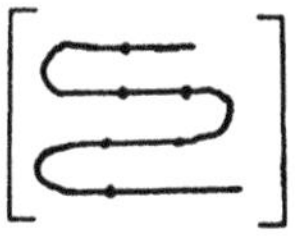
Epidermal Growth Factor	40 ± 2	3	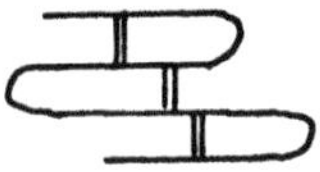
Fibronectin, Type I	45 ± 2	2	
Fibronectin, Type II	60 ± 2	2	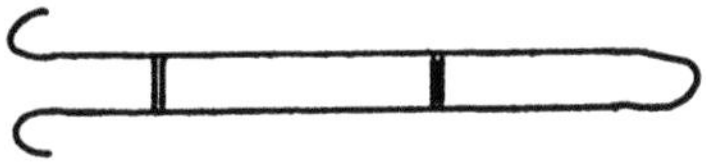
Proprotease "Kringle"	80 ± 2	3	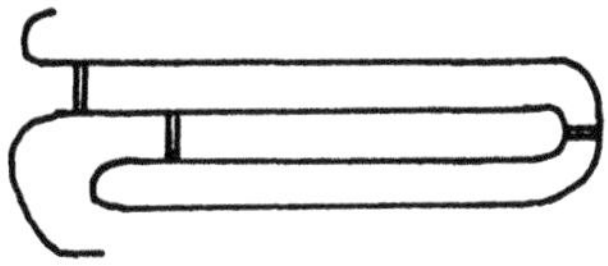

FIGURE 2 Some frequently exchanged modular units found in vertebrate proteins (from Doolittle[10]). The disulfide arrangement in the C9-type unit is not known.

TABLE 3 Some Proteins Known to Contain the EGF-Domain[1]

Thyroid Peroxidase	Thrombomodulin
Complement C9	Protein Z
Factor VII	Cartilage Matrix Protein
Factor IX	Vaccinia Virus
Factor X	LDL Receptor
Factor XII	Sea Urchin Regulatory Protein
Protein C	EGF-Precursor
Protein S	*Drosophila* Notch protein
Urokinase	*C. elegans* Lin-12
Plasminogen Activator	

[1] The EGF-unit, first observed in epidermal growth factor (EGF), consists of a segment 40–45 residues long containing 6 cysteines involved in 3 disulfide bonds.

HOW MANY PROTEIN FAMILIES?

Everyone knows that, given 20 amino acids arranged in strings of 300, the number of theoretical protein sequences possible is a superastronomical 20^{300}. It has dawned slowly that nowhere near this number of sequences has ever or will ever exist. Rather, a small number of starter types has been expanded by the general mechanism of duplication and subsequent modification, with a still-to-be-determined amount of intergenic shuffling. How many starter types there may have been is still a matter of conjecture,[5,7,30] but the sequence databases are now getting large enough that some limits can be imposed. Thus, it can be estimated that a search of any newly determined sequence has about an even chance of retrieving a related sequence in the data base. This is a very gross approximation, of course, for some of the relationships will reflect species differences and others only partial resemblances from shuffled exons. Furthermore, there is a great bias in the kinds of sequences being reported because of trends in biomedical research and funding. Still, the message is clear; the number of sequence types is not unlimited (Figure 3).

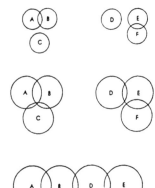

FIGURE 3   Ven diagram showing how the evolutionary network may come to include distant relatives as more sequences are reported. Thus, although the similarity between two distantly related sequences may be blurred beyond recognition, the resemblances of each to a third may reveal the relationship (from Doolittle[7]).

THE HUMAN GENOME INITIATIVE

The realization that all of biology is based on an enormous redundancy has extraordinary implications for the sequencing of the human genome. For one thing, it tells us that much of the genome will be immediately recognizable on the basis of sequence comparisons with existing databases, limited as they may seem to be. The other side of this bright coin is that sequencing the human genome is going to tell us a good deal about how organisms evolve in general. Thus, it is estimated that the human genome contains about a hundred thousand genes; as many as half of these may code for genes involved with the central nervous system. Seen another way, at least half of these genes have been assembled by duplicative expansion since the divergence of animals and plants, and as such, ought to have sequences that will be readily recognized one to another. We can count on there being a thousand kinases[16] and a thousand phosphatases, hordes of serine proteases and a corresponding army of serine protease inhibitors, a fundamental array of pharmacologic receptors[4] on the one hand, and another of the self-nonself variety[17,28] on the other. Beyond that, there will be a set of fundamental metabolic enzymes, perhaps a few thousand, that have changed at a relatively slow pace during the past two billion years and are still very similar in prokaryotes and eukaryotes.[12]

   All of these considerations have led me to make some predictions about the initial screening of sequence data that will come from the human genome project. First, even allowing for the problem of having to distinguish coding sequences from non-coding, at least half of the gene products in the human germline will be recognizably related to known proteins (or structural RNA genes). A second set of genes will be seen as belonging to families, but the families themselves will not yet have been identified. Of course, a small fraction of the sequences will have changed so much that their historical connections will be blurred.

The great sequence redundancy that we can anticipate will have another unexpected benefit: it should prove possible to model the three-dimensional structures of many of the inferred proteins on the basis of known X-ray structure of related proteins. All in all, then, the prospects for enlightenment afforded by the human genome project are highly favorable because of what sequence databases have already taught us.

ACKNOWLEDGMENT

Our computer studies on protein sequence comparison have been supported by NIH grant GM-34434.

REFERENCES

1. Barnett, D. R., T. H. Lee, and B. H. Bowman. "Amino Acid Sequence of the Human Haptoglobin β Chain. Amino- and Carboxyl-Terminal Sequences." *Biochemistry* **11** (1972):1189–1194.
2. Brew, K., T. C. Vanaman, and R. L. Hill. "Comparison of the Amino Acid Sequence of Bovine α-Lactalbumin and Hens Eggwhite Lysozyme." *J. Biol. Chem.* **242** (1967):3747–3749.
3. Brown, J. P., R. M. Hewick, I. Hellström, K. E. Hellström, R. F. Doolittle, and W. J. Dreyer. "Human Melanoma-Associated Antigen p97 Is Structurally and Functionally Related to Transferrin." *Nature* **296** (1982):171–173.
4. Bunzow, J. R., H. H. M. VanTol, D. K. Grandy, P. Albert, J. Salon, M. Christie, C. A. Machida, K. A. Neve, and O. Civelli. "Cloning and Expression of a Rat D_2 Dopamine Receptor cDNA." *Nature* **336** (1988):783–787.
5. Dayhoff, M. O. *Atlas of Protein Sequence & Structure.* Vol. 5, supp. 3. Washington, D.C.: National Biomedical Research Foundation, 1978.
6. Dixon, R. A. F., B. K. Kobilka, D. J. Strader, J. L. Benovic, H. G. Dohlman, T. Frielle, M. A. Bolanowski, C. D. Bennett, E. Rands, R. E. Diehl, R. A. Mumford, E. E. Slater, I. S. Sigal, M. G. Caron, R. J. Lefkowitz, and C. D. Strader. "Cloning of the Gene and cDNA for Mammalian β-Adrenergic Receptor and Homology with Rhodopsin." *Nature* **321** (1986):75–79.
7. Doolittle, R. F. "Similar Amino Acid Sequences: Change or Common Ancestry?" *Science* **214** (1981):149–159.
8. Doolittle, R. F. "Angiotensinogen Is Related to the Antitrypsin-Antithrombin-Ovalbumin Family." *Science* **222**:417–419.
9. Doolittle, R. F., M. W. Hunkapiller, L. E. Hood, K. C. Robbins, S. G. Devare, S. A. Aaronson, and H. N. Antoniades. "Simian Sarcoma Virus *onc* Gene, v-*sis*, Is Derived from the Gene (or Genes) Encoding a Platelet-Derived Growth Factor." *Science* **221** (1983):275–277.
10. Doolittle, R. F. "The Genealogy of Some Recently Evolved Vertebrate Proteins." *Trends in Biochem. Sci.* **10** (1985):233–237.
11. Doolittle, R. F. *Of URFs and ORFs.* Mill Valley, CA: University Science Books, 1986.
12. Doolittle, R. F., D. F. Feng, M. S. Johnson, and M. A. McClure. "Relationship of Human Protein Sequences to Those of Other Organisms." *Cold Spring Harbor Symposium on Quantitative Biology* **51** (1986):447–455.
13. Doolittle, R. F. "Redundancies in Protein Sequences." In *Prediction of Protein Structure & the Principles of Protein Conformation*, edited by G. Fasman. New York: Plenum Publ., in press.
14. Downward, J., Y. Yarden, E. Mayes, G. Scrace, N. Totty, P. Stockwell, A. Ullrich, J. Schlessinger, and M. D. Waterfield. "Close Similarity of Epidermal Growth Factor Receptor and v-erb-B Oncogene Protein Sequences." *Nature* **307** (1984):521–527.

15. Eck, R. V., and M. O. Dayhoff. *Atlas of Protein Sequence and Structure*. Silver Spring, MD: National Biomedical Research Foundation, 1965.
16. Hanks, S. K., A. M. Quinn, and T. Hunter. "The Protein Kinase Family: Conserved Features and Deduced Phylogeny of the Catalytic Domains." *Science* **241** (1988):42–52.
17. Hayashida, H., K.-I. Kuma, and T. Miyata. "Immunoglobulin-Like Sequences in the Extracellular Domains of Proto-Oncogene *fms* and Platelet-Derived Growth Factor Receptor." *Proc. Japan Acad.* **64** Ser. B (1988):113–118.
18. Hunt, L. T., and M. O. Dayhoff. "A Surprising New Protein Superfamily Containing Ovalbumin, Antithrombin-III, and Alpha$_1$-Proteinase Inhibitor." *Biochem. Biophys. Res. Commun.* **95** (1980):864–871.
19. Julius, D., A. B. MacDermott, R. Axel, and T. M. Jessell. "Molecular Characterization of a Functional cDNA Encoding the Serotonin 1c Receptor." *Science* **241** (1988):558–564.
20. Kobilka, K. B., H. Matsui, T. S. Kobilka, T. L. Yang-Feng, U. Francke, M. G. Caron, R. J. Lefkowitz, and J. W. Regan. "Cloning, Sequencing, and Expression of the Gene Coding for the Human Platelet α_2-Adrenergic Receptor." *Science* **238** (1987):650–656.
21. Kubo, T., K. Fukuda, A. Mikami, A. Maeda, H. Takahashi, M. Mishina, T. Haga, K. Haga, A. Ichiyama, K. Kangawa, M. Kojima, H. Matsuo, T. Hirose, and S. Numa. "Cloning, Sequencing and Expression of Complementary DNA Encoding the Muscarinic Acetylcholine Receptor." *Nature* **323** (1986):411–416.
22. March, L., and I. Herskowitz. "STE2 Protein of *Saccharomyces kluyveri* Is A Member of the Rhodopsin/β-Adrenergic Receptor Family and Is Responsible for Recognition of the Peptide Ligand αfactor." *Proc. Natl. Acad. Sci. USA* **85** (1988):3855–3859.
23. Masu, Y., K. Nakayama, H. Tamaki, Y. Harada, M. Kuno, and S. Nakanishi. "cDNA Cloning of Bovine Substance-K Receptor through Oocyte Expression System." *Nature* **329** (1987):836–838.
24. Rose, T. M., G. D. Plowman, D. B. Teplow, W. J. Dreyer, K. E. Hellström, and J. P. Brown. "Primary Structure of the Human Melanoma-Associated Antigen p97 (Melanotransferrin) Deduced from the mRNA Sequence." *Proc. Natl. Acad. Sci. USA* **83** (1986):1261–1265.
25. Vogt, P. K., T. J. Bos, and R. F. Doolittle. "Homology between the DNA-Binding Domain of the *GCN4* Regulatory Protein of Yeast and the Carboxy-Terminal Region of a Protein Coded for by the *onc* Gene *jun*." *Proc. Natl. Acad. Sci. USA* **84** (1987):3316–3319.
26. Waterfield, M. D., G. T. Scrace, N. Whittle, P. Stroobant, A. Johnsson, A. Wasteson, B. Westermark, C.-H. Heldin, J. S. Huang, and T. F. Deuel. "Platelet-Derived Growth Factor Is Structurally Related to the Putative Transforming Protein p28 Virus." *Nature* **304** (1983):35–39.
27. Weinberger, C., S. M. Hollenberg, M. G. Rosenfeld, and R. M. Evans. "Domain Structure of Human Glucocorticoid Receptor and Its Relationship to the v-*erb*-A Oncogene Product." *Nature* **318** (1985):670–672.

28. Williams, A. F. "A Year in the Life of the Immunoglobulin Superfamily. " *Immunology Today* **8** (1987):298–303.
29. Young, D., G. Waitches, C. Birchmeier, O. Fasano, and M. Wigler. "Isolation and Characterization of a New Cellular Oncogene Encoding a Protein with Multiple Potential Transmembrane Domains." *Cell* **45** (1986):711–719.
30. Zuckerlandl, E. "The Appearance of New Structures and Functions in Proteins during Evolution." *J. Mol. Evol.* **7** (1975):1–57.

Databases

Christian Burks
Theoretical Biology and Biophysics (T-10), Theoretical Division, Los Alamos National
Laboratory, Los Alamos, NM 87545

The Flow of Nucleotide Sequence Data into Data Banks: Role and Impact of Large-Scale Sequencing Projects

A number of individual labs or consortia are setting up (and many more are planned) large-scale sequencing efforts based on highly programmatic and/or automated approaches to determining nucleotide sequences. This development has raised a number of issues regarding the flow of nucleotide sequence data from these projects into the data banks that now serve the molecular biological community: Will journal articles still serve as the primary forum for explicit presentation (and transmission) of nucleotide sequence data? If not, how will the data get into the data banks? How will these changes in data flow affect strategies for storing and manipulating nucleotide sequence data? These issues are examined and a general strategy offered. In addition, a number of unresolved issues are discussed.

INTRODUCTION

The GenBank,[6] DNA Data Bank of Japan,[13] and EMBL Data Library[8] nucleotide sequence data banks currently contain about 25 million nucleotides spread across about 22 thousand sequences. Most of these data were incorporated into the data banks through (i) visual scanning by data bank staff of journals known to contain significant numbers of sequence reporting articles, followed by (ii) manual entry by data bank staff of the sequence data into the database.

As the rate of new data being published[5] several years ago began to exceed the in-house staff resources of the data banks, they began to explore alternative means of collecting the data. The strategies explored focussed on shifting the scanning and initial entry tasks from the centralized staff to the community generating the data. Thus, the role of the data bank staff was seen as shifting from that of providing centralized, manual entry to that of providing protocols and software tools that allow those generating new data to send automatically processable submissions to GenBank.[10]

Due in large part to the cooperation of a number of forward-thinking journal editors who have incorporated submission of new data to the data banks into the publication processes for their journals, the data banks have been able to make significant strides toward this new model for data collection. GenBank, for example, is now gathering roughly half the data it is responsible for through direct submissions from the scientists generating the data. The overall strategy[6] and specific mechanisms[7,9] currently supporting direct submission of data have been described elsewhere.

Though there is still room for improvement, the move to shift to direct, computer-readable submissions of data associated with journal publications is well on track. On the other hand, there is a current need to examine the mechanism for interaction between the data banks and the large-scale nucleotide sequencing efforts that are now getting off the ground.[2,15] Though these projects may not have technically come of age,[1,17] it is clear that many such projects are already in the planning—and even implementation—stage. As they are being described in planning discussions and funding proposals, they often raise new questions (relative to the character of data previously limited to the context of publication in journal articles) about the timing and frequency of data transmission to the data banks, proprietary concerns, and data representation in the database.

Approaches to addressing these questions should be addressed by not only the data bank staffs, but also by large-scale sequencing project staff, software developers, and the developers of large-scale sequencing stations. We have previously presented[3,5,6] preliminary views on this topic, focussing especially on the impact of greatly increased streams of data on data bank operations. Here, we focus on the characteristics of large-scale sequencing projects and their strategies for both managing their data and transmitting it to the data banks, and focus also on data transmission and representation issues that the projects and the data banks will have to address jointly.

LARGE-SCALE SEQUENCING PROJECTS
WHAT IS A LARGE-SCALE SEQUENCING PROJECT?

Though there is no established definition, there currently appear to be two criteria for the projects being identified as such: they plan to sequence larger regions (e.g., complete genomes) than have been sequenced in the past, and they plan to carry out the sequencing systematically and completely, often with a highly programmatic and/or automated approach, to greatly improve the efficiency of sequencing (e.g., number of nucleotides per person-year).

The first criterion alone does not completely distinguish these projects from the labs that have traditionally sequenced large amounts of DNA in the process of answering particular molecular biological research questions. And the second criterion alone does not distinguish the large-scale sequencing projects from the sequencing resource stations or groups that many institutions have set up to provide access to automatic sequencers. The large-scale sequencing projects are driven by biology, in the sense that the the region being sequenced represents an interrelated set of objects (structural or functional) that will further knowledge in a specific domain of molecular biology. On the other hand, the resulting sequence data (and most often the sequencing strategy) will not represent a set of data radiating out from one or more loci of predetermined interest; rather, the sequence data will represent a complete, contiguous representation of the information (most of which may be ill-understood in terms of biological function) from one end of the region to the other.

DETAILED CHARACTERIZATION

The following discussion broadly characterizes the large-scale sequencing projects now being planned or implemented. There are of course exceptions to the descriptions given, but the projects being implemented have a great deal in common.

Because these projects will generate such prodigious amounts of information, they are all being planned with the assumption that sequence data will be managed on computers. This will of course provide an excellent platform for transmitting data from these projects to the data banks, but only if the local software, hardware, and networking environments have been well planned and carefully developed.

For the same reason, most of the data from these projects will not appear explicitly in the pages of a journal article, though in many cases the data will be *summarized* in journal articles. This trend is already apparent in journal policies for much smaller sequences. Thus, primary access to the sequence data will be through the data banks.

Currently, the data banks enter and contain more experimentally determined (or confirmed) annotation than annotation derived solely from pattern searches and comparisons. However, as indicated above, sequences will increasingly be determined that contain significant, and even a majority of, regions for which biological function(s) have not yet been experimentally identified or confirmed. Much of the annotation (delineation of boundaries of regions of specific functionality) will—at least initially—be the result of computer analyses and comparisons to similar but better-characterized data sets. It has always been the case that sequence or annotation data are occasionally revised in light of subsequent analysis or experimental work, but these revisions and additions, occuring after the sequence data are generally available, will occur with much greater frequency in the future.

In most cases, data will not be held back until the individual project is complete; as significant sub-regions are compiled and analyzed, they will be released to the data banks.

How much data is going to be coming out of these projects? Of those projects currently being planned or implemented, the length of DNA being contemplated is most often in the 10^6–10^7 nucleotide range, with the data being generated over a 3–5 year period at minimum. The DNA being focussed on ranges from complete bacterial genomes, to complete chromosomes in small eukaryotic genomes, to large sub-regions of individual chromosomes of the human genome.

In summary, these projects will be generating and managing nucleotide sequences one or more orders of magnitude larger than has previously been the case; as a consequence, they are using computer-based tools for managing, analyzing, and transmitting their data. These data will not appear in journal articles but will be—following direct submission—publicly accessible primarily through the data banks. The data will often be publicly accessible at a much earlier stage, with regard to their experimental characterization and consequent delineation of functionality, than the data traditionally found in the data banks.

DATA FLOW FROM SEQUENCING PROJECTS INTO DATA BANKS

GENERAL MODEL

Figure 1 presents a general model for describing the flow of data from large-scale sequencing projects into the data banks. Level I encompasses computers wired to the apparati (e.g., automated sequencers) used in generating sequence data; these data collection stations aid in interpreting digital data to create computer files containing sequence data, and may have some data management and/or analysis capabilities. Level II represents the stage at which data from one or more data-generation stations are compiled and merged (and conflicts recorded and/or resolved) onto computers in project data centers. These centers also support sequence analysis (e.g., searches for similarities to other sequences and other pattern recognition algorithms that lead to putative functional assignments for the sequences). The general data distribution occurs in Level III, where data bank computers are used to collect all submitted nucleotide sequence data and provide distribution of these data to the general scientific community.

The model encompasses those cases where Level I and Level II computing are done on the same individual computer (see node #1 in Level I, Figure 1) and also small projects where a single sequencing station is feeding data into a larger computer for data analysis (see node #2 in Level I and node C in Level II, Figure 1). The "standard" situation is envisioned to be the case (see nodes #3–4 in Level I and node D in Level II, Figure 1) where data from several data collection stations—possibly located at geographically distinct sites—are fed into a single project data center. In very large projects, there may be primary and secondary project data centers (see nodes A and B in Level II, Figure 1); normally, one of these centers would assume the role of transmitting the compiled data for the entire project to the data banks.

ROLE OF PROJECT DATA CENTERS

Those designing and setting up the project data centers should discuss their projects with data bank staff as early in the planning stage as possible; they should alert the data bank staff to the kind and amount of sequence data that will be appearing, and they will, in turn, acquire information about protocols and software that will facilitate the transmission of data from the centers to the data banks. This is also the best time to identify (and resolve) any disparity between the data items and structures proper to the individual project and those supported by the data banks. As the data are accumulated, the data banks will benefit from being kept closely apprised of the state of development of the overall data so that they can revise (or provide revision tools to the project data center staff) the data already in the data bank; for example, if the projects convene workshops to discuss, analyse, and synthesize the sequence and annotation data, it might be very useful to have data bank staff participate.

ROLE OF DATA BANKS

The primary role of the data banks will be to provide information and software tools that will support the transmission of data from the project data centers to the data banks. The data banks are currently restructuring the nucleotide sequence data in the context of relational database management systems and are also developing software to support "submit to data bank" stand-alone (or integrated) modules on remote systems.[6,9] Both of these efforts contain elements that may be of great use to those setting up local data management and transmission centers. In particular, the data banks can provide the detailed specifications for transactions coming into the data banks as transmissions from the project data centers. (These specifications will also be made available to independent software developers who are interested in developing their own "submit to data bank" modules.)

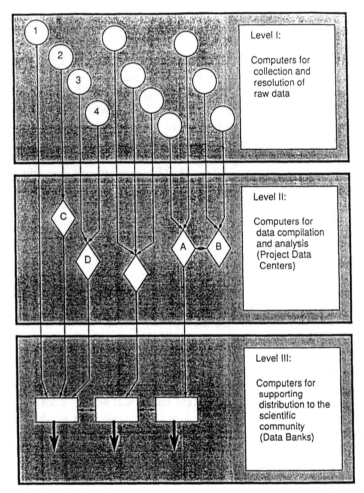

FIGURE 1 Data flow from large-scale sequencing projects into the data banks.
Computers in Level I are those wired to the apparati (e.g., automated sequencers) used
in generating sequence data; these stations aid in interpreting digital data to create
computer files containing sequence data and may have some data management and/or
analysis capabilities. Level II computers represent the stage at which data from one or
more data-generation stations are compiled and merged (and conflicts recorded and/or
resolved). These centers also support sequence analysis (e.g., searches for similarities
to other sequences and other pattern recognition algorithms that lead to putative
functional assignments for the sequences). Level III computers are those used by the
data banks (e.g., DDBJ, EMBL, and GenBank) that collect all nucleotide sequence data
and provide distribution of these data to the general scientific community.

ISSUES ARISING

The relative success of the model described above will depend on the progress made on a number of problems that are now being explored both by those setting up the projects and those responsible for maintaining the data banks. The following is a description of several problems that need to be addressed.

INTERPRETATION AND RESOLUTION OF DATA GENERATED AT SEQUENCING STATIONS

Those who have developed and are developing automated sequencing stations have inherited several difficult challenges with regard to providing software (and therefore greater automation, reproducibility, and speed) solutions to converting digital signals to sequences (including instances where signals are difficult to distinguish from background noise and instances where signals conflict), merging conflicting sequences, and assembling large numbers of sequences of varying degrees of overlap. Because large-scale sequencing projects will be transmitting data that have not been as thoroughly checked through independent experimental assessment of functionality, users of the data may want to have access to ambiguities regarding base assignments that are traditionally eliminated before the data are provided to the general scientific community. Similarly, maintaining detailed information on the degree of over-determination of individual regions may be very useful in assessing ambiguous assignments from different sources. Data structures allowing for the maintenance of this information in conjunction with the associated sequence data will be necessary.

SOFTWARE SUPPORTING DIRECT SUBMISSIONS

As mentioned above, the data banks are now developing software that will support entry of nucleotide sequence and annotation data at remote sites by the scientists generating the data.[6] Though in many cases this software may be directly applicable to the environment (and data flow) for individual projects, there may be instances where it makes more sense to either rewrite the front end on the software provided by the data bank or rewrite the entire software package. (In either case, these efforts should be guided by the specifications for transmission of database transactions provided by the data banks).[6,9] For example, the software currently being developed by the data banks is focussed on the individual investigator submitting data to the data banks on an occasional basis; the demands of submissions that could be occuring at one site daily (or more frequently) would make highly desirable the elimination of manual entry for each submission of information that may be identical (e.g., organism being sequenced) from submission to submission.

QUALITY OF DATA AND TIMING OF RELEASE OF DATA

Publication of nucleotide sequence data in journal articles has until now provided a straightforward framework for assessing when data should be released to the public (including one's competitors): data were published when one's peers (in the form of several referees) decided that (i) the data had been determined carefully and correctly, and (ii) the conclusions based on the data represented a significant step forward in science.

As discussed in the previous sections, nucleotide sequence data will increasingly be submitted directly to the data banks and often independently of (or prior to) journal publication.[4] This has raised many concerns about the quality of these data, and an interest in developing automatic (software-mediated) checks for the internal consistency of the data (e.g., do the protein-coding regions that are delineated agree with the sequence data?) as well as direct input from the community on the data in the data banks.

Of course, without publication in journals, omitting to submit data directly to the data banks will mean no access to the data at all. This concern is currently being addressed by the funding agencies supporting large-scale sequencing projects by delineating in the project specifications the timing for release of data to the data banks. Though this approach is readily implementable for contracts, it is not yet clear what mechanisms, formal or informal, will be brought to bear for data generated on research grants.

FREQUENCY AND MECHANISM OF REVISION OF DATA SETS

If, as discussed in previous sections, data in the data banks are going to require constant updating to reflect revisions and extensions provided in perhaps near-real-time by sequencing projects, data banks will have to be able to process these revisions much more automatically than is currently the case. Furthermore, the data banks (and perhaps individual data centers) will have to have a much more rigorous (and, again, highly automatic) method of keeping track of the change status of not only their own database but also those with which they strive to maintain identical copies. The former requirement is being addressed in the current restructuring of the data banks;[6,9] the latter concern has not yet been addressed as thoroughly.

THE HUMAN GENOME IS NOT A LINEAR ARRAY

Given that individual haploid human genome differs, on average, by 1 million or more nucleotides from other haploid human genomes,[11] an exact characterization of "the human genome" would have to include[12] all of the approximately 10 billion haploid genomes. In other words, the human genome is not a linear array but a 3.5×10^9 by 10.0×10^6 matrix. In less dramatic terms, it is important to note that data banks and analysis software have thus far favored the representation and manipulation of linear nucleotide sequence data. As large-scale sequencing projects evolve, though there may often be an initial emphasis on a single haploid sequence, interest will very quickly turn to orthogonal data on germline and somatic variations, especially those that can be linked to clinical phenotypes. Routine analyses and representation of these data will not only require a conceptual extension to two-dimensional data, but will also require an exact language for describing sequence variation that is richer than that available now.

Currently, sequence alignments are described on a nucleotide-by-nucleotide basis with "match," "mismatch," "insertion," and "deletion." Consider the alignments for the two following pairs of sequences:

aaaaaaaaaaaaaaaaaaaattttttttttttttttttttt

aaaaaaaaaaccccccccccgggggggggggttttttttttt

aaaaaaaaaaaaaaaaaaaattttttttttttttttttttt

aaaaaaaaaaattttttttttaaaaaaaaaaattttttttttt

Using the standard vocabulary, both of these alignments would be described by "$(\text{match})_{10}$–$(\text{mismatch})_{20}$–$(\text{match})_{10}$."

If, however, we recast these sequences in terms of regions rather than nucleotides,

(region 1) - (region 2) - (region 3) - (region 4)
(region 1) - (region 5) - (region 6) - (region 4)

(region 1) - (region 2) - (region 3) - (region 4)
(region 1) - (region 3) - (region 2) - (region 4)

we might choose to describe the first alignment as "$(\text{match})_{10}$–$(\text{mismatch})_{20}$–$(\text{match})_{10}$," as before, but describe the second as, "$(\text{match})_{10}$–$\text{flip}(10 \rightarrow 10)$–$(\text{match})_{10}$." The kind and quantity of mutational events implied by the description of the second alignment are quite different from the description of the first.

Thus, it would be useful to explore formal languages for describing and querying patterns in sequence *alignments* (including, for example, genetic rearrangements and other kinds of higher-order mutation), perhaps drawing on elements from formal pattern specificiation languages that have been developed for manipulating (e.g., Schroeder and Blattner[16]) and detecting patterns in (e.g., Myers[14]) nucleotide sequences.

ACKNOWLEDGMENTS

We would like to thank the participants in the workshop session on data flow from large-scale sequencing projects for providing insights into the projects they are setting up, and are indebted to members of the GenBank and Human Genome Information Resource projects for many fruitful discussions of the topics discussed here. In addition, conversations with several individuals have been particularly enlightening: G. Bell, D. Benton, G. Cameron, G. Church, M. Cinkosky, J.-M. Claverie, D. Cohen, A. Danchin, P. Gilna, J. Hayden, E. Hildebrand, J. Hochs, M. Hunkapillar, T. Hunkapillar, T. Marr, H. Mewes, S. Miyazawa, E. Myers, G. Olson, and L. Tomlinson. This work was funded in part by a contract (NO1-GM-7-2110) with the NIH and in part under the direct auspices of the U.S. Department of Energy.

REFERENCES

1. Alberts, B. M., D. Botstein, S. Brenner, C. Cantor, R. F. Doolittle, L. Hood, V. A. McKusick, D. Nathans, M. V. Olson, S. Orkin, L. E. Rosenberg, F. H. Ruddle, S. Tilghman, J. Tooze, and J. D. Watson. *Mapping and Sequencing the Human Genome.* Washington, D.C.: National Academy Press, 1988.
2. Bell, G. I.(1989), "The Human Genome: An Introduction." This volume.
3. Burks, C. "The GenBank Database and the Flow of Sequence Data for the Human Genome." In *Biotechnology and the Human Genome: Innovations and Impact,* edited by A. D. Woodhead and B. J. Barnhart. New York: Plenum Press, 1988, 51–56.
4. Burks, C. "Sources of Data in the GenBank Database." In *Biomolecular Data: A Resource in Transition,* edited by R. R. Caldwell, D. G. Swartz, and M. T. McDonald. Oxford: Oxford University Press, 1989, 327–334.
5. Burks, C. "How Much Sequence Data Will the Data Banks Be Processing in the Near Future?" In *Biomolecular Data: A Resource in Transition,* edited by R. R. Caldwell, D. G. Swartz, and M. T. McDonald. Oxford: Oxford University Press, 1989, 17–26.
6. Burks, C., M. J. Cinkosky, P. Gilna, J. E.-D. Hayden, Y. Abe, E. J. Atencio, S. Barnhouse, D. Benton, C. A. Buenafe, K. E. Cumella, D. B. Davison, D. B. Emmert, C. M. Etnier, M. J. Faulkner, J. W. Fickett, W. M. Fischer, M. Good, D. A. Horne, F. K. Houghton, P. M. Kelkar, T. A. Kelley, M. Kelly, M. A. King, B. J. Langan, J. T. Lauer, N. Lopez, C. Lynch, J. Lynch, J. B. Marchi, T. G. Marr, F. A. Martinez, M. J. McLeod, P. A. Medvick, S. K. Mishra, J. Moore, C. A. Munk, S. M. Mondragon, K. K. Nasseri, D. Nelson, W. Nelson, T. Nguyen, G. Reiss, J. Rice, J. Ryals, M. D. Salazar, B. L. Trujillo, L. J. Tomlinson, M. G. Weiner, F. J. Welch, S. E. Wiig, K. Yudin, and

L. B. Zins. "GenBank: Current Status and Future Directions." *Meth. Enzymol.*, 1989, in press.

7. Burks, C. and L. J. Tomlinson. "Submission of Data to GenBank." *Proc. Nat. Acad. Sci. USA* **86** (1989):408.

8. Cameron, G. "The EMBL Data Library," *Nucl. Acids Res.* **16** (1988):1865–1867.

9. Cinkosky, M. J., D. Nelson, J. W. Fickett, T. Marr, and C. Burks. "Overview of the New GenBank Architecture." Manuscript in preparation, 1989.

10. Fickett, J .W. "The Database as a Communication Medium." In *Biomolecular Data: A Resource in Transition*, edited by R. R. Caldwell, D. G. Swartz, and M. T. McDonald. Oxford: Oxford University Press, 1989, 295–302.

11. Kimura, M. "Gene Pool of Higher Organisms as a Product of Evolution." *Cold Spring Harb. Symp. Quant. Biol.* **38** (1974):515–524.

12. Kimura, M. "How Genes Evolve: A Population Geneticist's View." *Ann. Genet.* **19** (1976):151–168.

13. Miyazawa, S. "DNA Data Bank of Japan: Present Status and Future Plans." 1989, this volume.

14. Myers, E. "An Approach to Pattern Matching for Biosequences." Talk given at SFI workshop, December 1988, Santa Fe, NM.

15. Phillipson, L. "The DNA Data Libraries." *Nature* **332** (1988):676.

16. Schroeder, J. L. and F. R. Blattner. "Formal Description of a DNA Oriented Computer Language." *Nucl. Acids Res.* **10** (1982):69–84.

17. U.S. Congress, Office of Technology Assessment "Mapping Our Genes – The Genome Projects: How Big, How Fast?" Washington, D.C.: U.S. Gov't Printing Office, 1988.

Sanzo Miyazawa
Laboratory of Genetic Information Analysis, Center for Genetic Information Research, National Institute of Genetics, Mishima, Shizuoka 411, Japan

DNA Data Bank of Japan:
Present Status and Future Plans

Activities of the DNA Data Bank of Japan are reported; the present status of data collection, data entry and search/retrieval systems developed at the DDBJ, and future plans of the DDBJ are discussed.

INTRODUCTION

As a center for DNA sequence databanks and related activities in Japan, the DNA Data Bank of Japan (DDBJ) was established at the National Institute of Genetics with a grant from the Japanese government in April, 1986. The government has shown its support and commitment to the databank through a permanent grant. At present, this project consists of two faculty positions, some operating funds, and computer facilities.

A primary task of the DDBJ is, of course, DNA sequence collection. However, in addition, we have a wide range of activities: (1) DNA data collection and data entry in collaboration with other databanks; (2) data distribution, including the secondary distribution of the GenBank[2] and EMBL[3] databases in Japan; (3) provision of on-line access to DNA and related databases; (4) development of research tools for sequence analysis; (5) regularly published newsletters to inform researchers of the activities of the DDBJ, and (6) provision of training courses for users of the DDBJ computer system. We have been developing a data entry system to manage data collection, a search/retrieval system for sequence databases, and research tools for DNA and protein information analysis. To let researchers know about these activities, we have published newsletters regularly and given training courses. The newsletters contain articles that describe the state of international collaboration among databanks and how to submit data to databanks as well as matters of specific interest to Japanese scientists such as available databases at the DDBJ, how to access the DDBJ computer system, and how to use the databases. The activities of the DDBJ are also conveyed through an on-line service of the DDBJ computer system; information about our activities may be obtained by accessing the computer system and using the "getinfo"[10] command devised for this purpose. All these activities provided by the DDBJ are open to anyone regardless of whether one works for a non-profit organization or not.

In the following paper, I will briefly report the present status of data collection, data entry and search/retrieval systems developed at the DDBJ, and future plans of the DDBJ.

DATA COLLECTION

DNA data collection and data entry are the primary tasks of the DDBJ. Our data collection is carried out in collaboration with the GenBank[2] and the EMBL Data Library.[3] We started collecting DNA data in December, 1986. Data is currently entered in the GenBank format and fully annotated.

TABLE 1 The Number of Entries and Bases in Each Release of the DDBJ Database

Release	Date	Entries	Bases
1	07/87	66	108,970
2	01/88	142	199,392
3	07/88	230	345,850
4	01/89	302	535,985

We released the first version of the DDBJ database in July, 1987. Release 1 included only 66 entries and 108,970 bases. Since then, our database has been released every half year. The numbers of entries and bases included in each release are listed in Table 1. The DDBJ collected about 240,000 bases in one year from July 1987 to July 1988. About 8,000,000 bases were collected during the same period by the EMBL Data Library and the GenBank. In other words, the DDBJ processed about 1/30 of the total collection of DNA sequences in a year.

Each release includes a coding sequence database and a peptide sequence database that were extracted and translated from the original DNA sequence database. They are helpful for users, since the translation of base sequence to amino acid sequence is not trivial due to the exon-intron structure in a gene and the variation of genetic code. Release 2 and the later release included the journal index, accession number index, short directory, and data submission form files. Release 4 will be available in January, 1989.

JOURNALS SCANNED

One of international collaborations in data collection is to share journals that each databank scans. The DDBJ principally has charge of journals published in Japan. The journals scanned are listed in Table 2; the MEDLINE literature databases are used to search other journals.

As expected, original DNA sequences were hardly reported in most journals published in Japan except for a few such as the *Journal of Biochemistry (Tokyo)*, *Agricultural Biological Chemistry*, and the *Japan Journal of Genetics*. Even the *Journal of Biochemistry (Tokyo)* included only about 20–25 papers per year. The total number of papers that included original DNA sequences in the scanned Japanese journals was only about 30–40 per year. We are now planning to regularly scan a few main journals and use the MEDLINE literature databases to search minor journals in which few reports of original DNA sequences were published in a year. By the way, reports from Japanese research organizations[14] numbered 148 of the 1279 papers published in 1987 according to the BIOSIS Preview database. We processed about 70 papers in the past year, nearly half of the reports from Japanese organizations. This means that the DDBJ processed about 1/20 of all papers published in a year. (BIOSYS does not always cover all of the reports, so this number may be an overestimate.)

A main obstacle to increasing data entry is that it is difficult for us to employ enough annotators and reviewers. At present, we have only 0.5 full-time employees (FTEs) for annotation and 0.2 FTEs for review. It is reasonable to expect that the DDBJ could process only about 1/20 or 1/30 of the total collection if one compares the DDBJ staff with the staffs at the GenBank and the EMBL Data Library. Direct data submission from authors and even data entry by authors

TABLE 2 Journals Scanned by the DDBJ and the Number of Papers Found to Include Original DNA Sequences[1]

		entries	papers
JOURNALS PUBLISHED IN JAPAN			
Agricul Biol Chem	Vol. 50(01)–50(12) 1986	3	3
	Vol. 51(01)–51(12) 1987	12	11
	Vol. 52(01)–52(10) 1988	14	12
Cell Struc Funct	Vol. 11(01)–11(04) 1986	0	0
	Vol. 12(01)–12(04) 1987	0	0
	Vol. 13(01)–13(05) 1988	0	0
Chem Pharm Bull	Vol. 34(12)–34(12) 1986	0	0
	Vol. 35(01)–35(12) 1987	0	0
	Vol. 36(01)–36(10) 1988	0	0
Devel Growth Diff	Vol. 28(01)–28(06) 1986	0	0
	Vol. 29(01)–29(06) 1987	0	0
	Vol. 30(01)–30(04) 1988	0	0
J Biochem Tokyo	Vol. 99(01)–99(06) 1986	11	8
	Vol.100(01)-100(06) 1986	27	14
	Vol.101(01)-101(06) 1987	15	6
	Vol.102(01)-102(06) 1987	28	14
	Vol.103(01)-103(06) 1988	50	15
	Vol.104(01)-104(05) 1988	12	6
Jpn J Cancer Res	Vol. 77(01)–77(12) 1986	0	0
	Vol. 78(01)–78(12) 1987	1	1
	Vol. 79(01)–79(10) 1988	1	1
Jpn J Genet	Vol. 61(01)–61(06) 1986	10	2
	Vol. 62(01)–62(06) 1987	5	5
	Vol. 63(01)–63(05) 1988	1	1
Microbiol Immunol	Vol. 31(02)–31(12) 1987	3	2
	Vol. 32(01)–32(10) 1988	1	1
Plant Cell Physiol	Vol. 28(01)–28(08) 1987	2	2
Zool Sci	Vol. 3(01)–3(06) 1986	0	0
	Vol. 4(01)–4(06) 1987	0	0
	Vol. 5(01)–5(04) 1988	0	0
Nippon Ika Daigaku Zasshi[2]	Vol. 54 1987	2	2
JOURNALS PUBLISHED OUTSIDE OF JAPAN			
J Gen Virol	Vol. 68(03)-68(12) 1987	38	27
	Vol. 69(01)-69(11) 1988	53	33

[1] This data was collected in November 30, 1988.

[2] Not scanned.

themselves are absolutely necessary for us to further develop the DDBJ database. However, if one considers the significantly large amount of DNA sequence data that will be analyzed in near future and if one also wants to keep the quality of data annotation, obviously it is more practical and more realistic for any databank to encourage researchers to enter the data themselves rather than to increase staff.

In order to let authors submit data directly to the DDBJ, we made an agreement with some journals: a floppy diskette or a hard copy of the data submission form is sent to every author whose paper is accepted. We will try to increase the number of journals with which we have such an agreement and also to extend the relationship with all journals such that they do not accept papers without direct data submission similar to the agreement[6] between the EMBL Data Library and the Nucleic Acid Research. GenBank is developing software to help authors enter their data. At present, we plan to use their software.

DATA MANAGEMENT

Since the DDBJ computer system became available in April, 1987, we have been developing a data management system on our UNIX[13] system. Usually a database system consists of subsystems such as (1) a data entry system, (2) a search/retrieval system, and (3) a data analysis system. It would be desirable to manage all of the three systems by using a single management system. However, building such a system would take a time. We could not afford this method, because we already had started data entry and so decided to create each system independently. Not only is such a system easy to create, users can sort out the data entry system that they do not need.

Our computer system is connected to the JUNET network, which is a UUCP[15] network for electronic mail and bulletin boards in Japan. Researchers may send the DDBJ DNA sequence data by electronic mail or any media. If the journal in which that data is supposed to appear is not one of which the DDBJ has charge, we will forward it to an appropriate databank by electronic mail; electronic mail addresses have been established for the EMBL Data Library and the GenBank so data can be forwarded to them. The EMBL Data Library and the GenBank may also communicate with submitters through the DDBJ. A special login account is available for anyone to log onto and get a restricted access to the DDBJ computer[10] (see Figure 1). A primary purpose of this special account is to provide a way for researchers to obtain a submission form and to submit data to databanks.

```
niguts
     Welcome to the NIG, UNIX System V Release 2.0

login: DDBJnews
Terminal type (pc98msdos): vt100

          DDBJ online news

     available commands

menu              # type this menu list
getinfo           # get information
man               # get the manual of commands
mailx             # send a mail; "mailx ddbj < filename" for ddbj
addresses         # list e-mail addresses
ls                # list contents of directory
cat               # concatenate or type files
pg                # pager
cp                # copy files
rm                # remove files
vi                # vi editor
kermit            # file transfer program
conv              # remove <CR> and ctrl-z
exit              # exit

DDBJnews% getinfo
```

FIGURE 1   A special login account "DDBJnews" to provide a restricted access to the DDBJ computer system.

A DATA ENTRY SYSTEM

We built a data entry system by utilizing the Source Code Control System (SCCS)[15] available in a UNIX system. Each entry is managed as an SCCS file. SCCS has the following functions[16]: (1) version control: a record is kept with each set of changes, which includes what the changes were, why they were made, who made them, and when, and (2) file locking: only one person can modify data at a time. Both are useful in data entry where more than one person is working simultaneously.

Figure 2 shows data flow at the DDBJ and lists commands that are used at each step of the data entry. Direct data submissions both by floppy diskette and by electronic mail are managed by using mail queues. Data to be submitted is mailed to a special electronic mail address, DDBJsub. Databank staffs use the "accept" command to stamp an accession number on the mail and to forward it to a next mail address, DDBJacc. It can also send a return message of acknowledgement

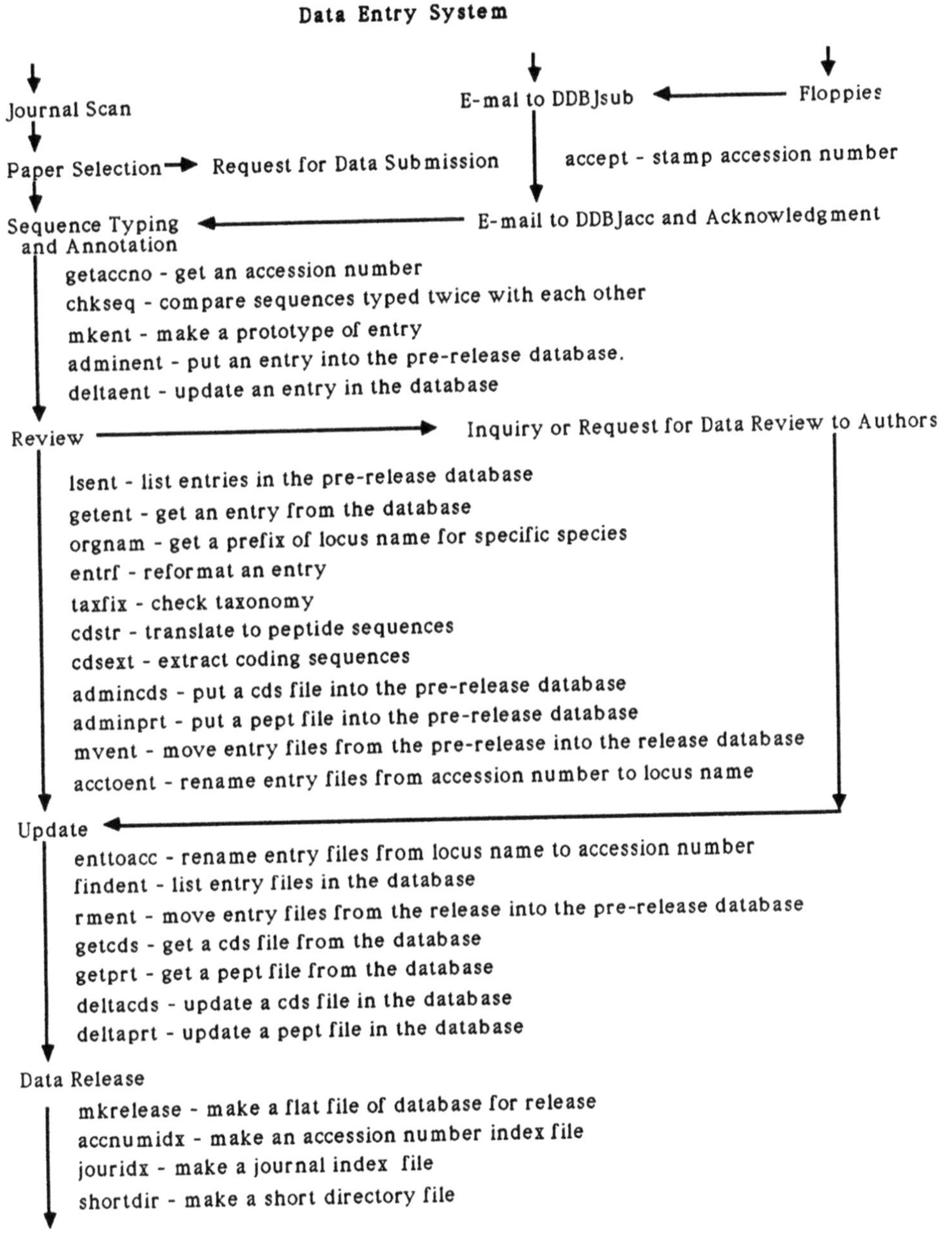

Data Entry System

```
    ↓                              ↓                         ↓
Journal Scan               E-mal to DDBJsub  ◄──────── Floppies

    ↓
Paper Selection ──► Request for Data Submission    accept - stamp accession number

    ↓
Sequence Typing  ◄──────────────── E-mail to DDBJacc and Acknowledgment
  and Annotation
      │  getaccno - get an accession number
      │  chkseq - compare sequences typed twice with each other
      │  mkent - make a prototype of entry
      │  adminent - put an entry into the pre-release database.
      │  deltaent - update an entry in the database
      ▼
Review ──────────────────────────►  Inquiry or Request for Data Review to Authors
      │
      │  lsent - list entries in the pre-release database
      │  getent - get an entry from the database
      │  orgnam - get a prefix of locus name for specific species
      │  entrf - reformat an entry
      │  taxfix - check taxonomy
      │  cdstr - translate to peptide sequences
      │  cdsext - extract coding sequences
      │  admincds - put a cds file into the pre-release database
      │  adminprt - put a pept file into the pre-release database
      │  mvent - move entry files from the pre-release into the release database
      ▼  acctoent - rename entry files from accession number to locus name
Update ◄──────────────────────────────────────────────────
      │  enttoacc - rename entry files from locus name to accession number
      │  findent - list entry files in the database
      │  rment - move entry files from the release into the pre-release database
      │  getcds - get a cds file from the database
      │  getprt - get a pept file from the database
      │  deltacds - update a cds file in the database
      ▼  deltaprt - update a pept file in the database
Data Release
      │  mkrelease - make a flat file of database for release
      │  accnumidx - make an accession number index file
      │  jouridx - make a journal index file
      ▼  shortdir - make a short directory file
```

FIGURE 2   Data flow and commands that are used at each step of data entry.

to the submitter. Annotators process the mail that arrives at the DDBJacc address as well as papers that are found to include original sequence data during the journal scan. An entry is entered by using tools such as "chkseq" and "mkent," and then copied into a directory, which stores pre-release entries, by using a command "adminent." The "chkseq," which was programmed by Dr. H. Hayashida of the DDBJ, compares to each other the base sequences that are typed in twice, and the "mkent" makes a prototype of entry. The "adminent" and "deltaent" are front-ends for the "admin" and "delta" commands of SCCS, respectively; the "admin" creates an SCCS file and the "delta" is used to update it. Reviewers then get a pre-release entry by using the "getent" command, review it, and if necessary modify it. An entry may be updated by using the "deltaent" command. After data is reviewed and considered to be satisfactory, it is moved from that directory to another that stores entries ready for release.

QUALITY CONTROL. Data is checked in several ways. A DNA sequence is checked by typing each entry twice. Format, taxonomy, journal name, start and stop codons, and the codon frame in amino acid coding regions are checked by using programs such as entrf, taxfix, cdsext, and cdstr listed in Figure 2, which were programmed by Dr. J. Fickett of the GenBank. However, spelling in reference, features, and comment records is checked only by human review. We plan to use a spelling-checker for this portion. Our experience in checking only coding regions indicates that non-coding regions usually include errors. An effort to find errors in non-coding regions should be made.

A SEARCH/RETRIEVAL SYSTEM FOR SEQUENCE DATABASES: FLAT

We have been developing a search and retrieval system for flat file databases in order to provide simple tools to use DNA and protein sequence databases. This system called FLAT[11] consists of primitives, most of which perform a single operation and work as a filter in a UNIX system; a filter[8] program reads a line from standard input, processes it, and then writes some output onto standard output. Some basic commands available in the FLAT are listed in Table 3. They perform basic single operations such as (1) extraction of specified types of records from database files, (2) search of strings in each entry of database, and, if found, output of those entry names, (3) performance of "and," "or," and "xor" in respect of entry names, and (4) extraction of specified entries from database files. These filters may be combined with a UNIX pipe[8] to perform a complicated task. One may search and retrieve entries from databases by key words such as author name, journal name, title, organism name, source name, and any combination of such items. An example of such a search and retrieval request is given in Figure 3. This is a typical approach for designing programs in a UNIX system.

Strings for these programs are specified in the regular expression,[8,15] so that one can search and retrieve entries in databases by fuzzy key words and entry names. The "seqgrep" program also allows operators to use the regular expression to specify sequence patterns to be searched for in databases. Some of these filters were programmed in the Bourne shell and use UNIX tools such as sed, egrep, sort, and awk,[15] so that they are flexible enough to support many formats of databases and to easily keep up with the format changes that often occur. At present, the GenBank, EMBL, PIR[7] (Protein Identification Resource) and PRF[12] (Protein Research Foundation) data formats are supported. However, this approach tends to trade computational speed for flexibility. So applications whose processing speed is critical are written in C language. A program "getgb," which extracts specified entries from databases, uses a "pseud" index file to quickly find the location of the entries in a flat database file.

This FLAT[11] search/retrieval system for sequence databases is designed to be portable among UNIX systems that are available for a wide range of computers from super- to microcomputers.

TABLE 3 Some Basic Commands Available in Flat Software

{and | or | xor} *file1 file2* [*file3*...]
- and/or/xor entry names in *files*

{dirgb | dirembl | dirpir | dirprf} [*database-file*...]
- make short directory of *database-files*

{fromgb | fromembl | frompir | fromprf} [*file*...]
- convert *files* from the GenBank/EMBL/PIR/PRF format into the BIONET-like format

{getgb | getembl | getpir | getprf} [-1] [-0] "*database-files*" [*entry*...]
- get *entries* from *database-files*

{rcdgb | rcdembl | rcdpir | rcdprf} [-f "*database-files*"] *record-type*...
- get specific *record-types* from *database-files*

rsites *reg.-expr.-file* [*file*...]
- search sequence patterns specified in *reg.-expr.-file* in *files;* appropriate for search of restriction enzyme sites

seqgrep [-l *max-pattern-length*] *reg.-expr.* [*file*...]
- search sequence patterns of *full regular expression* in *files*

{srchgb | srcdembl | srcdpir | srcdprf} [*options-for-egrep*] *reg.-express.* [*database-file*...]
- search patterns of *full regular expression* in the text portion of *database-files*

```
niguts% flat
flat% set embl=annseq.dat
flat% rcdembl -f $embl OC | srchembl Vertebrata > vrt
flat% wc -l vrt
  8110 vrt
flat% rcdembl -f $embl DE KW RT | srchembl -i oncogene > onco
flat% wc -l onco
  537 onco
flat% and onco vrt > onco+vrt
flat% xor onco onco+vrt > onco-vrt
flat% wc -l onco+vrt
  386 onco+vrt
flat% wc -l onco-vrt
  151 onco-vrt
flat% getembl $embl < onco+vrt > onco+vrt.seq
flat% exit
niguts%
```

FIGURE 3   An example of search and retrieval by using FLAT software.

A GET-INFORMATION COMMAND: GETINFO

An online help program called "getinfo"[10] has been devised to provide databank staffs and users an easy way to access necessary information. One may use the "getinfo" to learn how to submit DNA data and to which databank it should be submitted; the user can even get a data submission form. An example of the "getinfo" command is shown in Figure 4.

The "getinfo" command apparently mimics the help utility of the VAX/VMS system.[16] However, unlike the VMS help utility, each item of information is stored as a flat file and organized into a tree-like structure, if necessary, by using symbolic links[15] or "pseud" symbolic links; the "pseud" symbolic link was devised because the symbolic link is not available in the System V UNIX. The "getinfo" displays a specified item and, if available, a list of help items at the next level and prompts the user to choose one of them. A "pager" program, "pg" or "less," available in a UNIX system, is used to print files on terminals so that one may read a help item page by page and may save it into another file, if necessary.

```
DDBJnews% getinfo
```
Type the name of item in which you are interested. The item will be displayed.
If you type
      <CR>, "getinfo" will back up to more recent topic.
      ctrl-c, "getinfo" will quit at that point.
      '?',   "getinfo" will output an item list again.
Meta characters for file names in "csh" may be used to specify an item.
      ex. "ddbj*", "*LAN"

Pager "jpg" is used to output files; to get help, type ": h" or "(page .): h".

DDBJ_news->	Info_as_file	Learn_unix	Learn_vms
Welcome_msg	background_job	bugs/	bulletin_board/
emacs/	file_transfer/graphics_lib/		imsl_stat_math/
ingres->	inquiries	junet/	line_printer
local_commands/	mails/	manuals->	nig_system/
printing_man	tex	troff	tty_emulator/
work_directory			

Item, <CR>, ctrl-c or ? **DDBJ**

DDBJ news

Application/	data_submit/	db_catalog	db_growth/
db_manuals/	db_version	dir_of_files	iris_softwares
manuals/	newsletters/	softwares/	vms_softwares

Item, <CR>, ctrl-c or ? **data_submit**

FIGURE 4 An example of using the "getinfo" command.

FUTURE PLANS
DATA COLLECTION

The DDBJ aims to process at least all of the DNA sequence data analyzed in Japan. However, at present, three DNA databanks—the EMBL Data Library, the GenBank, and the DDBJ—collaborate by sharing journals for scanning rather than sharing data entry on the basis of the geographic location of submitters. This method of cooperation is used because of a technical problem concerning data collection of regularly scanned journals. The system of data collection will change, as direct submissions from authors increase. Until then, the DDBJ will collect sequence data analyzed in Japan, but will forward them to the databank in charge without processing. We believe that this is a step toward the DDBJ taking charge of the collection of data analyzed in Japan.

DNA sequence data that we collect is sent to the EMBL Data Library and the GenBank when the DDBJ database is released every half year. They incorporate it into their databases. So the delay between data submission and its appearance in the EMBL and GenBank databases may be significant. In order to improve this situation, we plan to send the data as soon as it is ready for release. This new system will begin early in 1989.

COMPUTER NETWORK

The JUNET network to which the DDBJ computer system is connected is a UUCP[15] network that is connected by public telephone line and so it is a flexible but slow means of communication. Communication over telephone lines via modem is not stable. A current project of building and maintaining identical databases at three sites—the GenBank, the EMBL Data Library, and the DDBJ—needs high-speed communication among their computers. The DDBJ has a plan to connect the DDBJ computer system to the Internet in the U.S.A. to make high-speed communication feasible. Also, the DDBJ is planning to network computers of related organizations in Japan with the DDBJ computer system by using a X.25 packet communication line. This network would be useful in data collection and in DDBJ's provision of database access.

E. COLI SEQUENCING PROJECT

Because of significant breakthroughs, the large amount of DNA sequences of various genes in a wide range of organisms from prokaryote to human have been analyzed. Now, sequencing the entire genome is not just a dream but a feasible project.

In 1987, Kohara et al.[9] constructed the physical map of the whole *Esherichia coli* (*E. coli*) genome by isolating 3,400 lambda phage clones that contain segments of *E. coli* chromosome and by constructing a restriction map for eight 6-base-recognizing enzymes. Those clones, which may be used for the isolation of any desired *E. coli* genes, are maintained at the Laboratory of Gene Library in the National Institute of Genetics in Japan, and distributed to anyone who wants to use them. Kohara et al.'s work is the first case of the complete physical mapping of the whole genome. A project of sequencing the entire *E. coli* genome in Japan will use those *E. coli* clones. Directed by Dr. T. Yura and Dr. K. Isono, this project will start in April, 1989, under the direction of Dr. T. Yura and Dr. K. Isono. At present, about 15 laboratories are involved and more laboratories will join later. The DDBJ will join this project in managing sequence data.

    E. coli is one of those organisms whose genetics has been best studied at the molecular level. Its genome size is about 4.7 million bases and is presumed to consist of about 3000 to 4000 genes. About 15-20% of its base sequence and about 1000 genes[1] are known at present. The genome size of *E. coli*, 4.7 Mb, is not very long but only about one fifth of the total bases that have been collected by databanks. Even so, the physical mapping and sequencing of its entire genome and the management of the sequence data are not trivial tasks. (1) New types of information such as absolute and relative map positions and overlaps between sequence segments must be managed by a database management system (DBMS). (2) Researchers would certainly like to perform more diversified and complicated retrieval of sequence segments than what is currently available, such as retrieval of sequence segments by gene name, product name, and map position. DBMS must be flexible enough to satisfy these needs. (3) In addition,, the project involves many researchers in more than 10 laboratories who are located at geographically distant places. Even though the genome size is not long, the DDBJ cannot afford to enter all the data by itself, and so sequence data including annotation must be processed by each researcher. That is, distributed data entry must be solved. Data entry by authors is not a problem peculiar to this project but a general task which databanks must promote, as already mentioned. In order to make distributed data entry feasible, a user-friendly program for author entry is needed.

    The problems that I listed above would be common in any large-scale sequencing project. Of course, the difficulty of a project would depend on the characteristics of the genome that one wants to analyze. The longer the genome size is, the more efficient the experimental and computer systems are required to be. Also, the existence of highly repetitive sequences in eukaryotes would make the physical mapping of their genomes difficult. The genome of *E. coli* would be easier to analyze and so it may be a good exercise for genome sequencing.

    A new feature table definition[5] has been just completed by the EMBL and the GenBank with the assistance of the DDBJ in September, 1988, in order to represent in proper form a wide range of new information on DNA sequences that have been discovered in the recent development of molecular biology. A relational database whose schema[4] has been designed by the GenBank may be flexible enough to manage a large quantity of sequence data with the proper representation of information required for large-scale sequencing. Such a database would also make it feasible to search and retrieve sequence segments in various ways. The DDBJ will prepare for this new trend in DNA database reconstruction.

ACKNOWLEDGMENTS

I would like to thank the GenBank, especially Dr. James W. Fickett, for kindly
providing us useful programs and data for managing the DDBJ database. I would
like to acknowledge Dr. Hidenori Hayashida, for his efforts in managing data review,
and also the staff of the DDBJ.

REFERENCES

1. Bachmann, Barbara J. "Linkage Map of *Esherichia coli* K-12, Edition 7."
   Microbiol. Rev. **47** (1983):180–230.
2. Burks, Christian. This volume.
3. Cameron, Graham N. "The EMBL Data Library." *Nucl. Acids Res.* **16**
   (1988):1865–1867.
4. Cinkosky, Michael J., Debra Nelson, and Thomas G. Marr. *A Technical
   Overview of the GenBank/HGIR Database.* Los Alamos, NM: GenBank/HGIR.
5. DDBJ, EMBL Data Library, and GenBank. "The DDBJ/EMBL/GENBANK
   Feature Table: Definition, version 1." Mishima, Japan: DDBJ; Heidelberg,
   FRG: EMBL Data Library; Mountain View, CA: GenBank, IntelliGenetics,
   1988.
6. EMBL and GenBank staffs. "A New System for Direct Submission of Data to
   the Nucleotide Sequence Databases." *Nucl. Acids Res.* **15(18)** (1987).
7. George, David. Personal communication, 1987.
8. Kernigan, Brian W., and Rob Pike. *The UNIX Programming Environment.*
   New Jersey: Prentice-Hall Inc., 1984.
9. Kohara, Yuji, Kiyotaka Akiyama, and Isono Katsumi. "The Physical Map
   of the Whole *E. coli* Chromosome: Application of a New Strategy for Rapid
   Analysis and Sorting of a Large Genomic Library." *Cell* **50** (1987):495–508.
10. Miyazawa, Sanzo. *A Guide to the DDBJ Computer System.* Mishima, Japan:
    DDBJ, National Institute of Genetics, 1987.
11. Miyazawa, Sanzo. *The Manual of the FLAT Database and Sequence Analysis
    System for DNA and Proteins.* Mishima, Japan: DDBJ, National Institute of
    Genetics, 1988.
12. PRF Peptide Sequence Database. Maintained by the Peptide Institute, Pro-
    tein Research Foundation, 4-1-2 Ina, Minoh, Osaka, Japan.
13. Ritchie, D., and K. Thompson. "The UNIX Time Sharing System." *CACM*
    17 (1974):365–375.
14. Uchida, Hisao. Personal communication, 1988.
15. *UNIX User's Manual and Programmer's Manual.* Berkeley, CA: Computer
    Science Division, Univ. of California, 1984.

16. *VAX/VMS Command Manual.* Massachusetts: Digital Equipment Corporation, 1987.

David Kristofferson
BIONET, c/o IntelliGenetics, 700 E. El Camino Real, Mountain View, CA 94040; email:
kristoff@net.bio.net

BIONET: Status and Future Plans

INTRODUCTION

The BIONET National Computer Resource for Molecular Biology[3,6] has grown over the past five years to serve a community of almost 900 laboratories worldwide. Because of the need to service large numbers of geographically dispersed users who require access to genetic sequence data, BIONET has undertaken many initiatives in the area of computer networking and communications. These initiatives have immediate bearing on the proposed human genome project and are considered here in this context.

THE GENOME PROJECT

In a recent report entitled *Mapping and Sequencing the Human Genome*[4] the National Research Council raised many issues about the proposed genome project. Among the major points, I wish to consider two that have particular relevance to the work in progress at BIONET.

To quote from the report:

"The mapping and sequencing project will generate more data than any other single project in biology." (p. 75)

This raises the first issue of *how will the scientific community at large access and use this data.* Secondly:

"This project will also require an unprecedented sharing of materials among the laboratories involved ... access to all sequences and materials generated by these publicly funded projects should and even must be made freely available ..." (p. 8)

This obviously raises the next issue of *how can this "unprecedented" level of sharing be achieved.*

ACCESSING THE DATA

Currently, nucleic acid and protein sequence information is distributed by the major databases on magnetic tape and floppy diskette. The latter medium is already becoming strained. For example, release 57 of GenBank required 66 5.25-inch 360kb "XT"-style floppies or 22 1.2Mb "AT"-style floppies and the next release on floppies (version 59) will be about 20% larger.[1] Clearly the move is toward distribution of the databases on CD-ROM, but in the long run, given some of the projected rates of automated sequencing, this mechanism of distribution may also suffer from lack of timeliness compared to accessing the latest versions of the databases directly at the database production site via a computer network.

This last means of access may have several variations. Assuming that eventually most biologists at research institutions will have access to the databases via high speed (>1.5 Megabit per second) Internet-style network connections, one could foresee at least two possible means of accessing data. (1) Database access could take place by remote login over the network to the computer at the database maintenance site. This means that analyses would be run on the database organization's computer. This has the disadvantage of focusing all of the demand for computing power at one location (which can be overcome by providing additional hardware), but has the advantage of reducing the number of required service and support personnel compared to a more decentralized system. (2) Alternatively, a mechanism could be devised that would allow the database organization's computer to update automatically over the network the database copies at a number of regional computing sites. This has the advantage of distributing the computing demand more widely over existing hardware resources, but will require additional development work and technically competent support personnel to maintain the system at the various sites. A prototype of this latter system has been put into effect by the

EMBL in Europe,[7] but it is undoubtedly the case that further development work is needed before this system will be able to handle the anticipated data transfer rates.

Unfortunately, in the real world many research institutions are still limited to 9600 baud, essentially mail-only, BITNET connections that may be overwhelmed by the anticipated data transfer rates. Development work will be needed at each of these sites to upgrade hardware and network connections. BIONET is assisting several sites in developing the necessary network connections, and, longer term, is committed to implementing the networking possibilities described above.

In the near term BIONET is helping to ease the networking transition by providing a database searching service that can currently be accessed over any major network worldwide. Called "FASTA-MAIL," this service utilizes the FASTA program of William Pearson and David Lipman[5] for sequence similarity searching. BIONET has added an interface that allows a user on any accessible network to submit a query sequence by electronic mail to the BIONET computer and have the specified search performed automatically. Presently the GenBank and EMBL nucleic acid sequence databanks and the PIR and SWISS-PROT protein sequence databanks may be searched. The FASTA-MAIL program reads the mail input (without human intervention), performs the database search, and sends the results of the FASTA search including the top 20 sequence alignments back to the user by return mail. During weekends, the turnaround time from England to BIONET in California and back for a search of the PIR protein database (version 17) from submission of the initial mail message to the receipt of results was as little as 15 minutes.[2] This type of service is a concrete example of how, for the near future, widespread access to current database information can be made easily available. Of course, this system too will eventually be overburdened due to limitations on the amount of information that can be handled by mail systems, but it provides a near-term solution to the first genome issue raised above until the other proposed solutions can be implemented. The mail interface can also be modified to handle other types of data processing requests.

Access to FASTA-MAIL is freely available for non-profit research. The only requirement is that the intended user first send an electronic mail message to the address "fasta-req@net.bio.net" certifying that he/she is engaged in non-profit research. The user will then be registered to allow access to the program and instructions for using FASTA-MAIL will be sent by return mail.

ACHIEVING "UNPRECEDENTED" SHARING

As to the second genome issue raised above, computer networking also provides an ideal mechanism to allow the sharing of data and other information or technical

TABLE 1 BIOSCI Bulletins

BBOARD NAME	TOPIC
AGEING	Scientific Interest Group
BIONEWS	General announcements
BIOTECH	Biotechnology issues
BIO-CONVERSION	Scientific Interest Group
BIO-MATRIX	Applications of computers to biological databases
CONTRIBUTED-SOFTWARE	Information on public domain molecular biology programs
EMBL-DATABANK	Messages to and from the EMBL database staff
EMPLOYMENT	Job opportunities
GENBANK-BB	Messages to and from the GenBank database staff
GENE-EXPRESSION	Scientific Interest Group
GENOMIC-ORGANIZATION	Scientific Interest Group
METHODS-AND-REAGENTS	Requests for information and lab reagents
MOLECULAR-EVOLUTION	Scientific Interest Group
ONCOGENES	Scientific Interest Group
PC-COMMUNICATIONS	Information on communications software
PC-SOFTWARE	Information on PC-software for scientists
PIR	Messages to and from the PIR database staff
PLANT-MOLECULAR-BIOLOGY	Scientific Interest Group
PROTEIN-ANALYSIS	Scientific Interest Group
RESEARCH-NEWS	Research news of interest to the community
SCIENCE-RESOURCES	Information about funding agencies, etc.
SWISS-PROT	Messages to and from the SWISS-PROT database staff
YEAST-GENETICS	Scientific Interest Group

expertise. In March of 1987 BIONET encouraged the trend to on-line electronic submission of sequence data to the databases by releasing the "XGENPUB" program on the BIONET computer. XGENPUB was the first program to allow a user to annotate data and then directly submit it over the Internet to the GenBank and EMBL (and now PIR) sequence databases. Since the program was accessible on the same system where many researchers performed their sequence assembly projects, the data was not subject to possible errors of transcription, etc. XGENPUB helps get the data into the databases more rapidly and therefore

TABLE 2 Addresses for Information Regarding Participation in the BIOSCI Newsgroup Network

The Americas (Internet/BITNET/USENET):	biosci@net.bio.net
U.K. (JANET):	biosci@uk.ac.daresbury
Europe (EARN):	biosci@irlearn.ucd.ie
Scandinavia:	biosci@bmc.uu.se

provides faster access to the information by the community at large. On-line data
annotation and submission by the authors will be critical if the databases are to
handle the expected data flow.

Sharing can also be promoted through the use of electronic communications.
In 1984 BIONET began a series of bulletin boards dedicated to biological topics
(see Table 1). In the last four years this system has grown from one that was solely
available on the BIONET computer into a worldwide biological information network with four major news distribution sites covering all major international computer networks. Besides BIONET in the U.S.A., other participating sites in this
"BIOSCI" bulletin board service include the SERC Daresbury Laboratory in the
U.K., University College Dublin in Ireland, and the University of Uppsala in Swe-
den. In addition to having readers in North America and Europe, the bulletins are
distributed as widely as Australia, New Zealand, Japan, Taiwan, Korea, Israel, and
South America. Among the networks covered by the BIOSCI service are the Inter-
net/ARPANET/NSFnet (U.S.A.), BITNET (U.S.A. and elsewhere), NETNORTH
(Canada), EARN (continental Europe), JANET (U.K.), and USENET (worldwide).

As seen in Table 1, over twenty "newsgroups" are available. By sending a single electronic mail message to any of the groups, a scientist can communicate
simultaneously with all other participants in the group. This can promote the
sharing of experimental reagents (METHODS-AND-REAGENTS newsgroup) and
software (CONTRIBUTED-SOFTWARE, PC-SOFTWARE newsgroups), improve
communications between scientists and the databanks (GENBANK-BB, EMBL-
DATABANK, PIR, and SWISS-PROT newsgroups), improve communications between the funding agencies and scientists (SCIENCE-RESOURCES newsgroup),
promote collaborations between biologists and computer scientists (BIO-MATRIX
newsgroup), disseminate research results more rapidly (two journals, *Journal of
Bacteriology* and *CABIOS*, have begun distributing their Table of Contents in ad-
vance of publication on RESEARCH-NEWS, and others may follow soon), an-
nounce job openings (EMPLOYMENT), and finally improve communications in
individual research specialties as evidenced by the numerous research newsgroups.
Use of these facilities has been almost doubling each of the last few years as an increasing number of biologists learn how to use electronic mail and discover the great
utility of these services. It is not an exaggeration in the slightest to say that these

of facilities bring new meaning to the word "community" and can indeed allow the "unprecedented" level of sharing envisioned by the National Research Council.

Information on subscribing to or participating in the BIOSCI newsgroup network can be obtained by sending electronic mail to one of the addresses in Table 2. Please choose the site that is most conveniently located.

ACKNOWLEDGMENTS

The BIONET Resource is funded by a grant (#P41RR01685) from the Division of Research Resources, National Institutes of Health. BIONET would like to acknowledge the kindness of Dr. William Pearson in making the FASTA program available to the BIONET (and now worldwide!) community. I would also like to thank Mr. Eliot Lear, BIONET systems programmer, and Mr. Spencer Yeh, BIONET Applications Analyst, for their efforts in developing FASTA-MAIL.

REFERENCES

1. Benton, David, of GenBank. Personal communication.
2. Ginsburg, M. Personal communication.
3. Kristofferson, D. "The BIONET Electronic Network." *Nature* **325** (1987):555–556.
4. National Research Council. *Mapping and Sequencing the Human Genome.* Washington, D.C.: National Academy Press, 1988.
5. Pearson, W. R., and D. J. Lipman. "Improved Tools for Biological Sequence Comparison." *Proc. Natl. Acad. Sci. USA* **85** (1988): 2444–2448.
6. Roode, D., R. Liebschutz, S. Maulik, T. Friedemann, D. Benton, and D. Kristofferson. "New Developments at BIONET." *Nucl. Acids Res.* **16**,:1857–1859.
7. Stoehr, Peter, of EMBL. Personal communication.

J. Claiborne Stephens,†‡ Iva H. Cohen,† and Kenneth K. Kidd†‡
†Yale-Howard Hughes Medical Institute Human Gene Mapping Library, 25 Science Park, New Haven, Connecticut 06511 and ‡Department of Human Genetics, Yale University School of Medicine, New Haven, Connecticut 06510

The Human Gene Mapping Library: Present Status and Future Directions

HISTORICAL OVERVIEW

As with many other databases in biology and medicine, the databases maintained by the Human Gene Mapping Library (HGML) began in the laboratory of a senior investigator. In particular, two present HGML databases, LITerature and MAP, had their origin in 1973 as simple, indexed files in the laboratory of Professor Frank H. Ruddle of Yale University. Not coincidentally, this date was also that of the first international Human Gene Mapping Workshop (HGM1), held in New Haven. In 1981, the National Institutes of General Medical Sciences funded a proposal to design and implement a human gene mapping database under a computerized database management system (DBMS) that would allow for an anticipated moderate growth of data. The initial version was implemented in QBE, a relational DBMS sold by IBM. However, QBE was abandoned in 1983 because of high costs for search and retrieval, and other problems such as poor handling of text. Our current database management system, SPIRES, was chosen at that time because it provided: (1) fast, inexpensive searching capability; (2) low-cost uploading and index building; and (3) excellent text-handling capabilities. The Howard Hughes Medical Institute (HHMI) began funding the HGML in 1985. The timing coincided with the beginning of the ongoing tremendous growth in both the amount and complexity of relevant human data. As an example, Figure 1 shows the growth in the RFLP database since its inception following HGM6 through the recent interim Human Gene Mapping Workshop (HGM9.5, New Haven, August 1988). The three years of HHMI support of the HGML have also seen the development and continuing improvement of on-line access via TELENET, a service currently provided free of charge, except for international telephone access charges.

The HGML will continue to use SPIRES through the upcoming Human Gene Mapping Workshop (HGM10), for which the HGML databases will be a primary resource. However, we are now collaborating with several other groups to design a relational model for the data in the HGML. It is hoped that a single international model will be developed to improve the portability and flexibility of the databases. The first relational implementation of the HGML data will probably be in ORACLE, the DBMS used by the ICRF (Imperial Cancer Research Fund), EMBL (European Molecular Biology Laboratory), and several other biomedically oriented databases. A permanent move into a relational DBMS following the HGM10 conference is anticipated, so that the HGML can better cope with the increasing growth and complexity of data. As it stands now, our databases are routinely redesigned

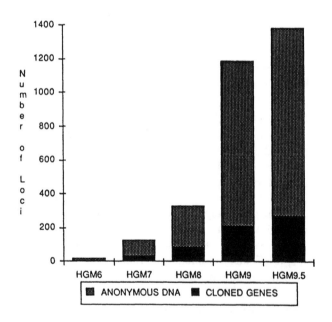

FIGURE 1 Trends in RFLP database size. HGM6, 7, 8, and 9 were held at two-year intervals. HGM9.5 was held one year after HGM 9.

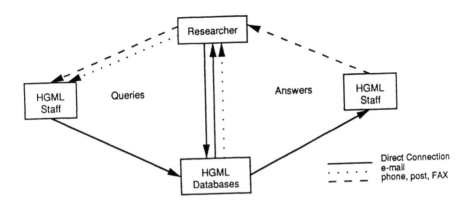

FIGURE 2   Modes of data retrieval and interactions with HGML.

as scientific directions shift in response to newer technologies. We view HGM10 as a particularly important testbed for new types of data and new representations.

CURRENT STATUS

ACCESS

There are many modes for accessing the data maintained by HGML, including direct connection via computer, electronic mail, and any means of communication with the HGML staff (Figure 2). Currently, on-line access involves an international community of several hundred researchers (Figure 3). This trend in on-line access represents a substantial growth, particularly over the last twelve months, to a current average of almost ninety different labs directly querying the databases a total of 350 to 400 times a month. These numbers do not take into account a considerable number of database searches done by the HGML staff on request.

CONTENT

The HGML currently maintains five interconnected databases (Figure 4). The MAP database is a catalogue of over 4500 known human genes and anonymous DNA segments that have been mapped to chromosomes or chromosomal regions. The RFLP database maintains detailed descriptions of almost 1500 loci that exhibit DNA

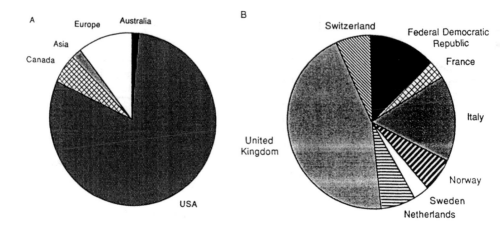

FIGURE 3 During 1988 the HGML databases were accessed 3489 times by 323 different individuals. A is Worldwide Access and B is European Access.

variation detected by restriction endonuclease analysis, including frequencies of the various alleles for diverse human populations, when known. The PROBE database maintains details on the size, vector, cloning, availability, and other characterizations of several thousand probes that have been useful in linkage or mapping studies. Our fourth database, CONTACT, is an on-line address book for researchers and others interested in human gene mapping and variation. Finally, the LIT database contains citations and other information about more than 10,000 literature references for entries in the other databases. Entries in MAP, RFLP, PROBE, and LIT are mutually cross-referenced to help the researcher obtain relevant information from more than one database simultaneously. Some important fields for each database are shown in Table 1. The data contained in these databases is retrievable via a system of menus appropriate for each database. A sixth database, MIM, is a version of Victor McKusick's *Mendelian Inheritance in Man*,[3] maintained by the Welch Medical Library at Johns Hopkins University and made available through collaboration between the Welch Library and HGML. Although the MIM database is not directly searchable through HGML, it is accessed via cross-references in MAP.

In concert with the increasing international realization of the need for networking the various biomedical databases,[1,4,5] the HGML has an ongoing effort to improve the level of cross-referencing to other databases. The HGML currently exchanges data with and cross-references OMIM (the electronic version of MIM mentioned earlier), GenBank, and the American Type Culture Collection (ATCC). Figure 5 shows, for each chromosome, the number of loci in the HGML for which there is relevant information in MIM (Figure 5A), the number of loci that are reciprocally cross-referenced with GenBank (Figure 5B), and the number of probes available or

TABLE 1 Sample data in HGML databases. Present databases MAP and RFLP will be generalized as LOCUS and POLYMORPHISM databases. Each entry in each database has a file number or unique identifier for internal and external cross-referencing.

LIT	LOCUS	PROBE	POLYMORPHISM	CONTACT
File #	File #	File #	File #	File #
HGM Symbol	HGM Symbol	HGM Symbol	HGM Symbol	Name
Author	Name	Lab Name	Name	Address
Citation	Map Location	Map Location	Map Location	City
Year	New Entry	New Entry	New Entry	State
Keywords	ARP Syntax[1]	ARP Syntax[1]	ARP Syntax[1]	Country
	Locus Type[2]	Locus Type[2]	Locus Type[2]	Phone #
	Sequenced (Y/N)	Molecular	Heterozygosity/PIC	
		Localization	(Information Content)	
	FAX #			
	Neighboring Loci		Probe Used	E-mail
		Enzyme Used		
	cross-referencing to other databases		Polymorphism Type	
Lodscores #	Enzyme Commission #	GenBank #	Method	
NLM #	MIM #	ATCC Access/	Population	
		Availability		

[1] <u>A</u>rbitrary <u>R</u>eference <u>P</u>oint localization of loci
[2] Gene or ananymous DNA

soon to be available from ATCC (Figure 5C), respectively. Figure 5C also shows the total number of probes in the HGML, the total number of loci in the HGML, and the number of these loci for which there is at least one probe in the ATCC collection.

The HGML databases have different primary sources of information and receive different degrees of monitoring by the scientific community. The LIT database draws primarily on scientific journals, article preprints, and proceedings and abstracts from various national and international conferences. Recent improvements in the data flow into LIT include agreements with journals such as *Genomics* and a collaboration with the National Library of Medicine (NLM). At present, editorial evaluation is primarily from the HGML staff in conjunction with HGML consultants.

The MAP database contains information from the literature and from the HGM Workshops. Although its contents were intended to represent the findings of the Human Gene Mapping Workshops, the two-year periodicity typical of these workshops required updating the MAP database in the interim, followed by a thorough revision at the conclusion of each workshop. Associated problems have been partially remedied by the active involvement of some of the HGM Committee chairpersons in editing the MAP database. This has been especially true for the Nomenclature and DNA committees. A more comprehensive plan being considered for the future is to allow the Chromosome Committee chairpersons to act as a standing board of editors as well.

The RFLP and PROBE databases rely on the literature and the HGM Workshops for most of their content, but also draw information from personal communications and from ATCC. The DNA Committee of the HGM Workshops works closely with HGML consultants and staff to maintain the integrity of these databases. HGML staff routinely interact with the authors of relevant literature to verify and complete the data extracted from the publications.

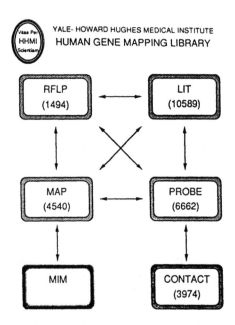

FIGURE 4   HGML database interconnections. Database sizes are as of late February, 1989.

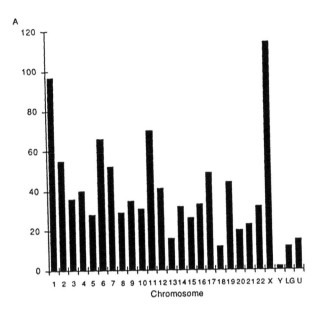

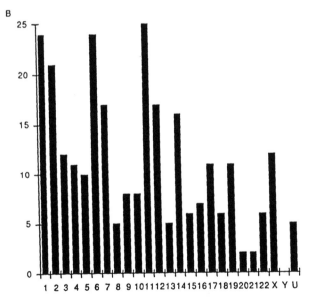

FIGURE 5 A: the MIM cross-references in HGML (by chromosome). As of December 1988, 1010 loci in HGML had cross-references to MIM. LG stands for "linkage group," as yet unassigned to a chromosome and U stands for "unassigned." B: the GenBank cross-references in HGML (by chromosome). As of January 1989, 271 loci in HGML were reciprocally cross-referenced with GenBank.

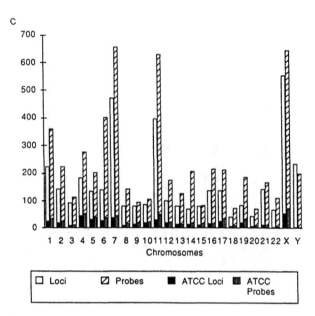

FIGURE 5   (continued) C: the ATCC cross-references in HGML (by chromosome). As of December 1988, 674 ATCC probes corresponding to 524 loci in HGML were reciprocally cross-referenced with ATCC.

PUBLIC EDUCATION

In addition to the on-line databases, the HGML has undertaken a number of other activities for disseminating data to the scientific community. Chief among these is the Chromosome Plotbook, published after each HGM Workshop and usually once in between workshops. The Plotbook is a graphical representation of each chromosome showing the chromosomal localizations of all mapped loci. This Plotbook is currently distributed free of charge. Also, the HGML has initiated a newsletter to provide tips, pointers, and interesting overviews to its user community. Figure 6 is an updated version of an overview graph that appeared in a recent newsletter.

The HGML has been involved in exhibits at several national and international conferences, serving the purposes of introducing the scientific community to the HGML as well as allowing direct feedback from the scientific community. HGML exhibits at these conferences generally provide on-line access, which serves multiple purposes as an on-line resource, a means of assisting novice users, and as an introduction to the HGML.

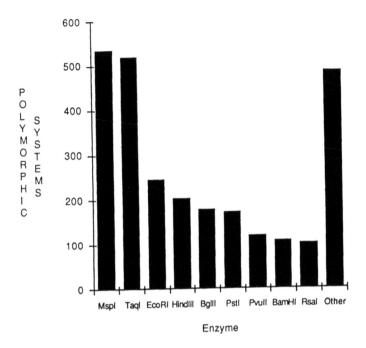

FIGURE 6   "Most popular" enzymes in the RFLP database—the distribution of
restriction enzymes used to discover 2666 polymorphic systems.

Several publications making use of HGML data have also appeared recently.
Papers by Track et al.[7,8] highlight the potential of the HGML's RFLP database for
the molecular biology and forensic communities, whereas one by Stephens et al.[6]
does the same for molecular evolution and population genetics. In addition, current
HGML database needs have motivated several collaborations between HGML staff
and the computer science and artificial intelligence communities.

PRESENT PROBLEMS, FUTURE PLANS

Maintaining a database that is of value to the scientific research community is an
ongoing challenge. We shall address this issue and the steps that are being taken
to meet this challenge.

USER FEEDBACK

A critical factor in maintaining a database service is input/feedback from the pool of potential users in the broad scientific community as well as from the existing community of users. The interactions noted earlier are providing considerable feedback, but additional measures are needed. Distributions of surveys, both written and on-line, are underway and planned to focus the attention of users on particular issues. General comments, suggestions, and criticisms are welcome at any time.

FLOW OF INFORMATION

A growing concern to many researchers is the frequent lack of complete documentation in many published descriptions of probes and RFLPs. This deficiency is largely the result of the quite heterogeneous backgrounds of the authors, the reviewers, and the journal editors involved in publishing relevant papers: many of these individuals are unaware of the needs and scientific standards of the other fields to which their work is relevant. Any remedy for this requires greater recognition by the scientific community of the value of having standard types of information accessible in an electronic database. Standardization of reporting greatly facilitates entry of these data into a database. For example, the *Nucleic Acids Research* RFLP Short Reports are a good model for standardization of reporting RFLP data. Even so, full papers in *Nucleic Acids Research* often abandon this format for reporting RFLPs and consequently many fail to provide "complete" information.

Along with other biomedical databases, we share the universal problem of interacting with diverse journals with their diverse standards and objectives. For example, the HGML cannot require a journal to publish extensive documentation of an RFLP if the discovery was incidental to the research reported in an article and the RFLP's existence is irrelevant to the focus of the journal. Even for journals with a molecular genetics orientation, not only are there few guidelines for the reporting of data (and none universally accepted), but there is seldom any pressure for the data to be communicated to the relevant databases. For HGML, *Nucleic Acids Research* and *Genomics* are the leaders in establishing standards for an expeditious data flow into the database, but these examples are unfortunately too few.

As journal space becomes too expensive to publish full descriptions of clones used to define RFLPs and of the RFLPs themselves, databases such as the HGML are well positioned to be the repository for these details that are so important to researchers wishing to use those clones and RFLPs. Thus, we see the HGML serving a role analogous to that of GenBank. For this to work, there needs to be a standard mode of recording data and submitting them to the HGML. Initial steps in this direction are underway with the ATCC: a common minimum subset of information has been agreed upon and a common submittal form prepared. An electronic version of the form is also being prepared.

VOLUME OF DATA

The anticipated influx of data associated with the Human Genome project requires that the data be entered electronically (i.e., not retyped) as much as possible. Electronic data submission is being used as part of the Tenth Human Gene Mapping Workshop (HGM10) which is interacting closely with the HGML staff and databases. The responses of the human genetic and molecular mapping communities to this new procedure and the computer-handling experience gained will provide useful guidelines for future developments along these lines.

COMPLEXITY OF DATA

Changes in the discipline are constantly calling for additional fields and different structures in our scientific databases. Since the scientific discipline is continuing to change rapidly, this ongoing redesign process is not likely to stop. We are currently redesigning each of our databases into models closer to the relational design, although we are continuing to use the existing hierarchical DBMS because of the demands of the upcoming HGM10 Workshop. It is likely that HGML data will undergo full conversion to a relational DBMS (e.g., ORACLE) soon after HGM10. We will also be using HGM10 as a testbed for many new types of data, representations, and modes of interaction with the data, and adopt them into HGML as appropriate.

EDITORIAL EVALUATION

Quality control and editorial review exist at several levels. The editorial task that is currently most onerous and time consuming is the preparation of the summaries in the MAP database. These are integrations of all published data into the best estimate of the position of the locus. It requires expertise in the various methods used as well as specific expertise on the relevant chromosomal region. The volume of data accumulating will not allow a small number of experts to provide these summaries in a timely fashion. Since the HGM Workshops are now annual meetings, this function is accomplished more efficiently by the committee chairpersons. A successful attempt was made at HGM9.5 to distribute the editorship of "workshop" versions of the HGML databases to the committee chairs themselves. This will be done again at HGM10, a full-fledged meeting of the Human Gene Mapping community, complete with abstracts and a virtual flood of new data. If these experiments of distributing the editorship continue to be successful, it is reasonable that we could integrate editorship by the standing HGMW committee chairs into the ongoing operations of HGML. Administratively, the Executive Committee of the Human Gene Mapping Workshops has already agreed in principle to move toward one global database using HGML as its basis.

LINKING RELEVANT INTERNATIONAL DATABASES

The HGML is but one of many international databases that must function as a unified resource for human genome mapping efforts. These include NLM, GenBank, EMBL, DDBJ (DNA database–Japan), PIR (Protein Information Resource), MIM, CEPH (Centre D'Etude du Polymorphisme Humain), Lodscores (B. Keats), ATCC, Restriction Enzyme database (R. Roberts), and others that are just being developed. Steps are already underway to link the HGML information unambiguously to several of these other databases.

Soon, every HGML LITerature entry for a full paper will include the unique identifier for the citation in the Medline database. Since commercial sources offer full text on-line for several journals and index these with the Medline unique identifier, this should be a useful aid to researchers searching the HGML. Reciprocally, the HGML LITerature unique identifier will be included as part of the Medline record, providing a convenient entry point from Medline into the HGML.

Procedures are currently being developed for the HGML Locus unique identifiers to be used in both the GenBank and CEPH databases. The HGML already includes GenBank accession numbers but the procedures being developed will assure more complete identification of all appropriate cross-indices. The CEPH database will also include appropriate pointers into the PROBE and RFLP databases. This will link the specific probes and marker systems used in the CEPH linkage studies with the description and definitions in the HGML.

Another set of reciprocal cross-references being developed is with the lod scores database maintained by B. Keats.[2] The reference numbers in that database will be linked to the HGML and Medline unique identifiers, and the HGML Locus unique identifiers will be associated with the symbols used in the lod scores tables. It is also possible that the HGML will provide on-line access to that database.

These linkages mentioned are but one way in which the HGML is trying to work toward the future coordination of databases relevant to the human genome project and biomedical research in general. Other relevant databases exist and other types of coordination will undoubtedly be needed. Each of the existing databases has developed to meet the needs of a particular set of researchers. Those needs have now broadened in all cases. The existing databases must now determine which needs are best addressed by which database and which needs are met by none. By coordinating efforts to avoid duplication and to assure appropriate cross-referencing among databases, all databases can concentrate on their individual areas of expertise to assure the availability of the resource needed by the scientific community.

REFERENCES

1. Gruskin, K. D., and T. F. Smith. "Molecular Genetics and Computer Analyses." *CABIOS* **3** (1987):167–170.
2. Keats, B. J. B. *Linkage and Chromosome Mapping in Man.* Honolulu: University Press of Hawaii, 1981.
3. McKusick, V. A. *Mendelian Inheritance in Man*, 8th ed. Baltimore: The Johns Hopkins University Press, 1988.
4. McKusick, V. A., and F. H. Ruddle. "A New Discipline, a New Name, a New Journal." *Genomics* **1** (1987):1–2.
5. Morowitz, H. J., and T. F. Smith. *Report of the Matrix of Biological Knowledge Workshop.* Santa Fe, NM: Santa Fe Institute, 1987.
6. Stephens , J. C., R. K. Track, and K. K. Kidd. "The Human Gene Mapping Library: A Resource for Studies of Molecular Evolution, Population Genetics, and Comparative Mapping." *J. Cellular Biochem.* **13C** (1989):125.
7. Track R. K., F. C. Ricciuti, K. K. Kidd. "Information on DNA Polymorphisms in the Human Gene Mapping Library (HGML)." *Banbury Report* **32**: DNA Technology in Forensic Science, in press.
8. Track, R. K., F. C. Ricciuti, R. C. Doute, I. H. Cohen, K. K. Kidd, and F. H. Ruddle. "Human Gene Mapping Library: An On-Line Database on Human Genes and Anonymous DNA Segments Accessible by Telenet." *ICSU Short Reports* **9** (1989):156.

Sequence Comparisons

J. F. Collins† and **S. F. Reddaway‡**

†Biocomputing Research Unit, Department of Molecular Biology, Edinburgh University, Edinburgh, EH9 3JR, Scotland, U.K. and ‡Active Memory Technology Ltd., 65 Suttons Park Avenue, Reading, RG6 1AZ, U.K.

High-Efficiency Sequence Database Searching: Use of the Distributed Array Processor

Careful mapping of the sequence comparison algorithm described by Coulson, Collins, and Lyall[3] has provided on the AMT DAP 510 machine a high-speed method of searching for local protein sequence similarities in databases. The results have also demonstrated the value to the biologist of such exhaustive methods. To provide the searching tools that will be needed when the sequence databases have expanded, particularly the database of genetic sequence information, novel methods will be required to maintain an adequate search-and-retrieval capability with the most powerful computers.

Such a method that exploits the features of the DAP is described, whose performance should provide the basis for adequate searching even when the database has reached the size of the human genome, or 3×10^9 bases of genetic sequence.

INTRODUCTION

The impact of modern methods of sequence analysis in Molecular Biology has generated an environment in which computer-based methods of analysis form an integral
and vital part of the research process. Indeed, not only are these methods required
to do the 'housekeeping' of sequence gathering and collation, but there are whole
areas of enquiry which arise as research topics once the methods of sequence comparison and alignment have been developed. For example, the detection of common
themes in genes or proteins depends not only on showing that consistent patterns
can be detected, but also that these patterns do not occur so frequently in other
contexts that their diagnostic value becomes uncertain.

Many biological research groups lack the resources to search the collected sets
of sequences maintained as international resources. This task has attracted many
solutions that restrict the computation required (see the review by Collins and
Coulson[2]), but the adoption of such 'approximate' methods is accompanied by the
acknowledged risk that important sequence alignments have been overlooked.

Exhaustive methods which offer more likelihood of finding all good alignments
are therefore strongly to be preferred, even though more computation is required.
It is likely that high-performance machines running the best exhaustive algorithms
will form an essential resource at national or, if computing power becomes relatively
cheaper, at local levels. The efficient implementation of exhaustive searches on a
machine of appropriate architecture can now produce levels of performance likely
to sustain the increasing demand from biologists in the immediate future.

SIMD machines have significant advantages for the efficient mapping of these
comparison problems. Much of the computation can be accomplished in parallel
operations that exploit the full power of the multi-processor array. Two machines in
particular, the Distributed Array Processor (DAP) from Active Memory Technology
Inc. (AMT) and the Connection Machine CM-2 (Thinking Machines), can also
exploit bit-processing capabilities to work extensively with

- logical states parallelizing conditional operations usually set inside loops in
serial programs, and
- low-precision arithmetic, which is particularly appropriate for this database
searching application.

The performance of the 32K CM-2 program reported by Lander et al.,[5,6] which
includes micro-coded inner loops, and is claimed to be optimally coded, reaches
25 million matrix cell updates per second. The program saves one result per pairwise
comparison, or about 2000 per query protein used.

The performance we now report for the 1024-processor DAP 510-4 is 7 million
matrix cell updates per second, using a high-level language. This program also saves
16,000 results for subsequent analysis, whatever the size of database used. A 4096-
processor DAP 610 would give 28 million updates per second. We estimate that the
use of assembler code would improve these performances by a factor of about 3.
It appears that differences in architecture between the two machines offer a clear

advantage to the DAP, though implementation differences may also play a part in the better relative performance produced by the DAP 510.

THE ALGORITHM USED

Coulson et al.[3] developed an extension to the exhaustive search algorithm of Smith and Waterman.[8] This finds and reports all paths, in the 2-D matrix used for the comparison of query sequence against database sequence, which correspond to the highest scoring (arbitrarily large) set of alignments, subject to the constraints that only one path may be reported from each start coordinate and that the paths are non-intersecting.

A description of an implementation of this algorithm on an earlier Distributed Array Processor has been published by Coulson et al.[3] The present machine has more memory and a faster cycle time, and together with substantial improvements in the mapping and coding, we have dramatically improved the performance while still remaining in FORTRAN-PLUS, the high-level coding language for the DAP.

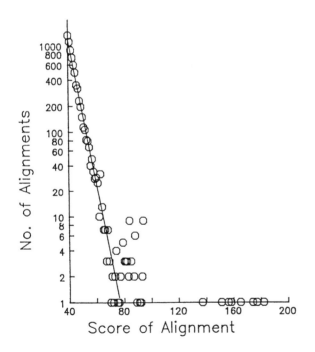

FIGURE 1 Distribution of the reported best alignment scores showing exponential decay. The fitted line defines the expected frequency of any score and thereby its significance.

```
.*.**.....* ** **  .  *
IHLUGQGISAHVAGAAGNKYT
VHVIGHSLGSHAAGEAGRR  T
```

FIGURE 2   Alignment of the amino acids (using single letter code) at the active site of the pig lipase with part of the Drosophila yolk protein. The essential serine (S) is substituted by a glycine (G).

EXAMPLES

Examples of sequence similarities discovered include the relationship of a cystic-fibrosis-associated antigen with mammalian calcium-binding regulatory proteins published by Dorin et al.[4] The interest here is that even if the strong alignment discovered had been absent, the cluster of proteins reported immediately below were all known to be calcium-binding proteins, and the biochemical nature of the new protein would have still been evident. The distribution of scores collected is shown in Figure 1, illustrating the behavior of the highest-scoring set as an exponentially decaying distribution, from which the expectation that an alignment of any specific score was expected with the particular database searched can be readily calculated.

Another interesting case is the similarity between part of the yolk proteins of *Drosophila melanogaster* and a triacylglycerol lipase from pigs discovered by Bownes et al.[1] Though the subsequent biochemistry has confirmed the role of this similarity in terms of a binding property of the yolk protein for acyl derivatives of the insect hormone ecdysteroid, the alignment of the region near the active site of the lipase (Figure 2) showed that the active serine required for lipase activity was absent. Searches based on the idea that similarities to enzymes are only of interest if the catalytic site has been maintained, would therefore have failed to attach particular importance to the local similarity, and might possibly have missed it altogether.

The use of this algorithm to collect a large set of results also allows significance to be reported. The method described by Collins et al.[2] requires no further database operations, nor randomization of the query sequence. The observed scores can then be given a rating relative to the pattern of alignments recorded between the bulk of the proteins in the database that are unrelated to the query sequence, and thus provide the biologist with a robust answer to the question of whether the similarity reported was likely to have occurred between unrelated proteins or whether it falls into that class of results that suggests a distinct degree of similarity that needs examination and understanding in biological terms.

DISCUSSION

The exhaustive search algorithm is extremely efficiently implemented on the Distributed Array Processor, and the large Arrays (64 × 64) machines will provide a further factor of 4 in increased performance, up to 28 million matrix cell updates per second. With even the current performance on the AMT DAP 510 machines, the database search becomes an experimental tool in its own right. With the appropriate set of parameters we can vary the selectivity and discrimination in a variety of ways, allowing the significance of individual alignments to be assessed not only in terms of improbability, but also in terms of whether they suggest biologically important hypotheses to the user.

The method of assessing the improbability of an alignment that has been described is robust and effective, removing the necessity of repeating searches with 'random' sequences to derive reference data on the distribution of alignments.

FUTURE COMPUTATIONAL NEEDS IN GENETIC SEQUENCE ANALYSIS

The future need will require major computational initiatives, as a consequence of

- the anticipated enormous expansion in the databases as major sequencing projects are funded, and
- the increasing knowledge and sophistication of the queries that should be made into the data.

The tasks will be to identify closely related sequences and to be able to find specific features within the database as rapid tests of hypotheses about significant subsequence organization, including signalling regions, splicing sites, etc.

The most effective searches must be geared to these objectives, and a natural method for the fast location of target regions may be the use of concordances ('dictionaries') of local sequences, which would allow the rapid retrieval of possible hits to the query sequence. The dictionary must be constructed in such a fashion that the likely search requests can be rapidly processed. Such retrieval systems have already been designed for text on the Distributed Array Processor by Reddaway and Page.[7] Multiple hashing techniques were described that were word based; they could be relied upon to produce a high degree of selectivity when reporting possible regions of hits, and were based on the generation of local hashing tables for individual segments of the text to be searched.

Nucleic acid databases pose a different problem, and we propose a variation of the scheme to retain the most important features of high-speed information retrieval. The use of local dictionaries seems eminently suitable. Each processor would examine the local dictionary available to it, with all processors acting in parallel and ranging through the query, in segments if necessary. The local counts of hits are tested to indicate whether further attention should be given to any particular region.

Nucleic acid sequences do not form natural words, and we must therefore ensure that the method chosen is not affected by casual displacements within the query or database sequences. We wish to take account of several specific features of nucleic acid data; viz.

- the alphabet is essentially confined to four symbols;
- sequences can be regarded as linear in storage and presentation even if in reality they may be circular; and
- stored sequences should not include high proportions of unknown or variant positions.

The scheme is to produce local dictionaries of single bits in an array, where a bit is set at the binary address in the array that corresponds to each n-base subsequence in the database segment being analyzed. Using a 2-bit representation of the four bases, a 5-base subsequence gives a 10-bit address into a table of 1024 bits. At least one table is stored under each processor and should have a low degree of occupancy to reduce the noise level of hits within a region of the query sequence being processed. The number of hits under each processor is accumulated and a test applied to decide whether within a specific block, or possibly between a pair of blocks, there are possible matches to significant parts of the query.

All words within the database sequences must therefore be entered into the local tables, and all words (i.e., starting from every position in the query sequence) then tested. By predefining the minimum number of hits (adjusted for the table bit density) required to initiate a future reference to a specific locality, a continuous series of searches, score accumulations, and tests will report all possible locations at which a significant threshold has been reached. The local table can be filled from a region of 256 bases, giving about 21% occupancy, after allowing for duplicates. This scheme is attractive, but an alternative scheme that might be preferable is outlined here.

The mapping of the database regions into local dictionaries should be partitioned into five separate tables, each table containing entries related to adjacent but non-overlapping 5-base sequences in the database. To keep the table density the same as in the previous case, the database region is increased to 1280 bases. Each query segment can be tested as five sets of non-overlapping sequences from the query sequence and the accumulated hit counts tested at appropriate intervals and in appropriate combinations from different tables, in order for the significant results to be stored and passed for further processing to a second stage of inspection. This should increase the discrimination possible between random levels of scores expected and the scores expected from runs of perfect or near-perfect matching bases.

Performance figures for the primary search can be given, and the use of processor arrays means that blocks of database sequence up to 1280 K bases can be scanned in one main operation. This data would occupy $1024 \times 5 \times 1024$ bits, or 625 KBytes. The execution of this search can be made extremely fast, and using a high-level language implementation on the DAP 510, we have measured a sequence comparison rate equivalent to 40,000 M character comparisons/second, which will allow effective database searching with query sequences of 10,000 bases against the database of 3,000,000,000 bases—the size of the human genome—in 750 seconds. Assembler coding increases performance by a factor of approximately 10, thus giving 400,000 M character comparisons/second, or 75 seconds for the primary scan, leading to a small amount of secondary sequence retrieval and display work. The DAP 610 would increase these rates by a further factor of 4.

In table dictionary form, the human genome database would occupy 1.5×10^9 bytes, or 0.5 bytes per base. For this to be transferred in 100 seconds, a transfer rate of 15 Mb/second is required. Fast disc systems can achieve this speed and will soon be available on AMT DAPs. In plain form, the database would be packed 4 bases to a byte; the corresponding transfer rate required is then 7.5 Mb/second. We will complete the implementation of this code on an AMT 510-8 1024 processor machine to assess the potential of dictionary-based database searching methods.

Extensions of these methods to alternative dictionaries based on three-base (6-bit) addresses would similarly allow scans for more local features and would be particularly apt for the scans of potential protein-coding regions. The shorter word length would reduce the speed by a factor of about 16.

Other variations of this technique can be readily envisaged.

CONCLUSIONS

The Distributed Array Processor shows promising performance on sequence database searches, provided care is taken to optimize the mapping of the computation onto the architecture of the machine. The current requirements of molecular biologists can be met, using the exhaustive search method,[3] thus providing reliable results to the user. There is no need to adopt less satisfactory approximate algorithms, which run the risk of failing to find important alignments.

The potential need for extremely high performance for searching the genetic databases can be met in principle by the remapping of the search as a locally based dictionary search, which offers useful response times even when databases of the size of the human genome (3×10^9 bases) are considered.

ACKNOWLEDGMENTS

J. F. C. gratefully acknowledges the support of the Darwin Trust of Edinburgh in this project.

REFERENCES

1. Bownes, M., A. Shirras, M. Blair, J. F. Collins, and A. F. W. Coulson. "Evidence that Insect Embryogenesis Is Regulated by Ecdysteroids Released from Yolk Proteins." *Proc. Natl. Acad. Sci. USA* **85** (1988):1554–1557.
2. Collins, J. F., and A. F. W. Coulson. "Molecular Sequence Comparison and Alignment." In *Nucleic Acid and Protein Sequence Analysis: A Practical Approach.* Oxford: I.R.L., 1987, 323–358.
3. Coulson, A. F. W., J. F. Collins, and A. Lyall. "Protein and Nucleic Acid Sequence Database Searching: A Suitable Case for Parallel Processing." *Computer J.* **30** (1987):420–424.
4. Dorin, J. R., M. Novak, R. E. Hill, D. J. H. Brock, D. S. Secher, and V. van Heyningen. "A Clue to the Basic Defect in Cystic Fibrosis from Cloning the CF Antigen Gene." *Nature* **326** (1987):614–617.
5. Lander, E., J. P. Mesirov, and W. Taylor. "Protein Sequence Comparison on a Data Parallel Computer." *Proceedings of the 1988 International Conference on Parallel Processing.* Philadelphia, PA: Penn. State Press, 1988, 257–263.
6. Lander, E., J. P. Mesirov, and W. Taylor. "Study of Protein Sequence Comparison Metrics on the Connection Machine CM-2." *Proceedings of Supercomputing* **2** (1989).
7. Reddaway, S. F., and R. M. R. Page. "High-Speed Data Searching with a Processor Array." *Microproc. & Microprog.* **24** (1988):655–660.
8. Smith, T., and M. S. Waterman. "Identification of Common Molecular Subsequences." *J. Mol. Biol.* **147** (1981):195–197.

Daniel B. Davison
Theoretical Biology and Biophysics (T-10), Theoretical Division, Los Alamos National
Laboratory, Los Alamos, NM 87545

Sequence Searching on Supercomputers

1. INTRODUCTION

Supercomputers allow the biologist to ask, and to answer, questions that would
otherwise be impractical or impossible. An example of current relevance is searching
the GenBank database with an entire HIV genome. The use of supercomputers in
sequence similarity searching at Los Alamos is not quite like working on similar
machines at, for example, an NSF-sponsored supercomputer center. There are a
large number of machines available, and two CRAY Y/MPs are coming on line
about now. The abundance of cycles, and the low cost, could lead one to think that
optimization would be less important. That is not so. Precisely because the queries
are larger and more involved, it is necessary to create code that is as efficient as
possible. This paper will discuss the steps involved in taking an existing similarity
code and improving its performance.

2. CFT AND CTSS EXPERIENCE

The CTSS operating system used at Los Alamos is unlike the CRAY COS operating system in that it does not support multitasking on multi-CPU CRAYs. However, other improvements are possible. The starting code was Minoru Kanehisa's SEQF, derived from code present in his IDEAS[1] package. That code was written fairly generally and seemed to be mostly devised for the DEC VAX architecture. The code did include two CTSS-specific techniques, one that computed the distances by antidiagonals, a technique originally suggested by Waterman.[4] The second uses a "feature" of CFT under CTSS; blank common is loaded last in the load image. By placing the library sequence in blank common, an "allocate more memory" system routine allows dynamic adjustment of memory for any size sequence. This can be advantageous in some situations and disadvantageous in others, as will be discussed below.

After using the code for some time, it became clear that with the increasing size of GenBank even supercomputer searches were becoming too expensive. The HIV genome run, for example, would have taken some 24 hours with the initial code. The CFT compiler used at LANL, version 1.13, contains a diagnostic code-flow-analysis routine called flowtrace. This measures the number of calls to each routine in the code and measures the time spent in each routine. This first analysis revealed ⅃hat the code was spending an inordinate amount of time in the subroutine which read in GenBank entries, specifically FORTRAN "format cracking." This operation is the conversion of data stored as characters on disk to a usable internal format. Table 1, row 1, shows the amount of time used by program operations. Preparation cf an "unformatted" version of GenBank and the appropriate changes to the I/O routines cut the total run time of the program by some 45% (Table 1, row 2). Curiously,

TABLE 1 SEQF Run Times With and Without Optimization

Routine Times[1]	Total Run	CPU	I/O	Mem
original	60.991	57.660	1.631	1.700
I/O opt.	28.133	23.065	3.545	1.524
% improvement	53.8%	60.0%	-117%	10%

[1] All times are in seconds. Tests were performed on an X-MP/48, query sequence of 188 nt, against the GenBank dnapri file, 6.52 MB, about 4 Mbases.

an expected performance improvement on the X/MP, placing the library in the solid state disk (SSD), did not change the run time. Testing confirmed that the library was placed in the SSD. The I/O bandwidth difference between CRAY disk and SSD implies that there should be a 12-fold improvement.

In order to improve the code further, hardware performance monitoring was done on the code by R. Koskela of C-3. That monitoring revealed that the dynamic sizing of the common block was causing a tremendous number of swaps in and out of memory, one for each sequence in GenBank. A simple modification, only increasing memory, improved run time slightly. This has interesting implications for performance of similar programs at NSF sites such as the National Center for Supercomputer Applications, which until recently ran CTSS on a X-MP/48. Those sites charge for runs by a time-and-space charge (memory usage × time occupied). This implies an uncomfortable tradeoff; one can either minimize the memory charges or I/O (for swapping) charges. I have not further investigated which would be most cost-effective. For those facilities, a combination of SSD usage and memory adjustment may be the best straightforward optimizing of similar codes.

The performance monitoring also revealed a surprisingly small percent vectorization, about 0.3%. The percent floating point operations was 100%. More sophisticated code monitoring than is possible with "flowtrace" or the LANL routine "TALLY" showed that there were three significant bottlenecks in the program. One, as would be expected, was the computation of distances by antidiagonals. There are two possible ways to vectorize this operation: by "conditional vector merges" and by gather/scatter operations (automatically performed by CFT 1.13). Testing revealed that CFT 1.13g (the latest CFT compiler certified by C Division at LANL) did not significantly affect the performance with either style. The other two locations were in sections of code that construct the k-tuple distances and build the traceback data structure.

The next performance enhancements involve better I/O management and multitasking. The I/O performance could be improved by not reading in each GenBank entry separately and instead concatenating all of GenBank (roughly 30 Mbases) into a single sequence. At three bits per nucleotide, this would require some 90 million bits (some 11 megabytes or 1.4 million 64-bit words). Asynchronous I/O, in an operating system that supports it, may improve the data input and result output times. The other obvious, and more usable choice, is to multitask the code.

Multitasking in this environment involves different considerations than on a truly parallel machine such as the Connection Machine. There are two types of tasking commonly available on CRAYs: microtasking and macrotasking. The former involves designation of short segments of code that can be executed asynchronously. In the latter, entire subroutines are executed in different processors. In this case it would appear that with the present code the most efficient use would be to have the n available processors running on m/n of the database's m bases on a X/MP. This would take maximum advantage of each processor's paths to and from memory for data transfer and assure independence of each processor's time. Much more sophisticated arrangements are possible, of course.

Another problem currently facing the CTSS programmer is the conversion of the simple arrangement of dynamic memory allocation routines in CFT to standard FORTRAN. Standard FORTRAN does not contain memory usage, but the CFT and CFT77 libraries, at least at Los Alamos, contain equivalent routines.

3. FASTA EXPERIENCES

Another sequence search code used at Los Alamos is the FASTA program, by Bill Pearson.[2] This code has been run on a CRAY 1S and an X-MP/14, both under UNICOS. UNICOS is the CRAY-hardware-specific version of Unix, based on System V Release 3.0. The performance of the unmodified code was very disappointing on the CRAY 1S. It typically took more than one wall-clock hour to receive one minute of CPU time. The rest of the time was spent in I/O waits. This appears to be due to two reasons: (1) the code dynamically allocates memory; each call results in a swap-out and swap-in. (2) The CRAY 1S has only one path from memory to CPU, so each reference to a load or store stops other activity. Typical run times were 12 to 16 hours for 12 to 14 minutes of CPU time. The CRAY X/MP-14, on the other hand, although a single processor, contains three memory-to-cpu paths, so execution times are better. The typical run time was about 6 hours to get 12 minutes of CPU time. This still is clearly unacceptable.

The performance of FASTA could be improved by the same procedures used in the SEQF program. If memory could be only taken and never given back, the majority of the swaps would be eliminated. Due to the structure of the code it is not clear how to accomplish this. The vectorization percentage is low; the CRAY C compiler vectorizes 2 loops in the original source code. Similarly, reading in more than one sequence at a time would eliminate considerable I/O. Additionally, buffering results in memory to eliminate the character- and line-at-a-time nature of the I/O operations would improve performance. Lastly, the use of a SSD under UNICOS with the new performance features (3) should be useful.

4. SUMMARY

The use of a supercomputer in computational molecular biology, and in sequence searching in particular, changes the nature of the questions that can be answered. At the same time, because of the cost and restricted availability of supercomputers, it is incumbent upon a researcher to make the most efficient use of such a resource. This article discussed experiences with two programs: SEQF, a CFT implementation of Wilbur- and Lipman[5]-type search code, and FASTA.[2] A number of optimizations have been made, with others pointed out. The guiding principle of these optimizations is to minimize the number of calls to system routines for memory allocation and I/O handling. A speed-up of some 45% overall was achieved for the FORTRAN code. Similar improvement can be expected for the FASTA code. Lastly, in terms of hardware used, it is clearly better to use a supercomputer with multiple data paths to and from memory, such as the CRAY X-MP and Y-MP series of processors, than the CRAY-1S or -2 series.

ACKNOWLEDGMENTS

This work was supported by an Alexander Holleander Distinguished Postdoctoral Fellowship award. In addition, NIH grant R01-GMS-37812 provided funding for computer time. This work was performed under the auspices of the U.S. Department of Energy.

REFERENCES

1. Kanehisa, M. "Los Alamos Package for Nucleic Acids and Proteins." *Nucleic Acids Research* **10** (1982):183–196.
2. Pearson, W., and D. J. Lipman. "Improved Tools for Biological Sequence Analysis." *Proc. Natl. Acad. Sci. USA* **85** (1988):2444–2448.
3. Reinhardt, S. "A Blueprint for the UNICOS Operating System." *CRAY Channels* **10** (1988):20–24.
4. Waterman, M. S. Personal communication.
5. Wilbur, W.J., and D. J. Lipman "Rapid Similarity Searches of Nucleic Acid and Protein Data Banks." *Proc. Natl. Acad. Sci. USA* **80** (1983):726–730.

Robert Jones,†‡ Washington Taylor IV,†* Xiru Zhang,** Jill P. Mesirov† and Eric Lander‡+
†Thinking Machines Corporation, 245 First Street, Cambridge, MA 02142-1214; ‡Whitehead Institute for Biomedical Research, 9 Cambridge Center, Cambridge, MA 02142; +Harvard University; *University of California, Berkeley, ** Brandeis University, Waltham, MA

Protein Sequence Comparison on the Connection Machine CM-2

INTRODUCTION

The comparison of a new sequence of a gene or protein against the database of known sequences has become a standard tool in the analysis of protein structure and function. Sequence similarity between two proteins may permit a function to be ascribed to an uncharacterized gene product, and study of the conserved regions between several related proteins may identify an important structural feature such as an enzyme active site. As the determination of sequence without prior knowledge of function becomes commonplace, as in projects to characterize entire genomes, inference from sequence similarity will become even more important.

Sequence comparison algorithms based on dynamic programming have emerged as the most sensitive for this task. Variants on the fundamental method permit the best matching subsequence,[11] or best N subsequences,[15] between two sequences to be identified or the best global alignment to be generated. These regions can contain mismatches, insertions or deletions permitting weak but biologically significant relationships between proteins to be identified. This level of sensitivity is achieved at the relatively high computational cost of order N^2 with respect to sequence length,

N, which can become excessive in searching a database where 10,000 or more comparisons are required. The rate at which new sequences are being generated serves to compound this problem.

Currently, depending on implementation, searching a database using dynamic programming can take from several hours to overnight on minicomputers, and this has led to faster but approximate methods based on hashing to be developed. However, these methods do not have the sensitivity of dynamic programming; weak matches, particularly those that require a large number of insertions and deletions, may be missed. As it is precisely this sort of weak relationship that provides the most insight into protein structure and function, much work has been devoted to improving the performance of the dynamic programming approach.

One way to achieve this is to perform the most expensive parts of the algorithm on a piece of specialized hardware, such as a custom VLSI chip, that is designed solely for this application and is therefore very fast. This approach has been proven in many other fields of computation, such as graphics or signal processing. However, it requires a high investment in the production of the hardware and sacrifices the flexibility of software for execution speed. A successful implementation of this approach has the potential to bring fast database searching directly to many research labs and is being actively pursued by several groups.[5,8]

Alternatively one can implement the algorithms on supercomputers. Vector machines such as the CRAY[TM] and parallel machines such as the Connection Machine CM-2[TM] and the Distributed Array Processor[TM] (DAP) can approach the speed of custom-made hardware but retain the flexibility of a software implementation of the algorithm. This is particularly important in a field in which the details of the algorithm may change with time, as has been the case in sequence comparison. Although the expense of supercomputers limits their availability, many are accessible to researchers as shared resources over nationwide networks such as the Arpanet.

As described below, sequence comparison is very well suited to implementation on a massively parallel computer, that is, one in which a large number of connected processing elements perform parts of the overall task simultaneously. For our application the effect is to reduce the running time from $O(N^2)$ on a serial computer to $O(N)$ on a parallel machine. Here we present our implementation of this algorithm on the data parallel Connection Machine CM-2,[6,7] manufactured by Thinking Machines Corporation. In their contribution to this volume Collins and Reddaway describe a different implementation on the DAP.

The CM-2[2,4] consists of up to 64K single-bit processing elements, each associated with 64K bits of local RAM. Special floating point hardware is available as an option, but is not required by this application. Processors communicate via a high-speed network that can be configured into N-dimensional grids; in this application we configure the processors as a one-dimensional array. The machine functions in SIMD (Single-Instruction, Multiple-Data) mode; that is, all processors execute the same instruction stream but apply this to their own local data. Programs for the CM-2 are written and executed on front-end computers, such as VAX[TM], Sun[TM] or Symbolic LISP Machines[TM], using parallel extensions to common programming languages, such as LISP, C or FORTRAN, and in PARIS, the parallel instruction set for the CM-2. Parallel instructions are passed to a sequencer on the CM-2 which then broadcasts appropriate low-level operations to the processors. One feature, unique to the CM-2 among SIMD machines, that is of particular value to this application is the ability to perform indirect addressing at the level of individual processors. This permits different memory locations to be accessed in response to a single instruction.

ALGORITHM AND IMPLEMENTATION

The appropriate algorithm for searching a database is that of Smith and Waterman,[11] which locates the best common subsequence between two sequences. This enables a relatively small conserved region between two otherwise unrelated proteins to be identified.

Briefly, for two sequences $A = (a_1, a_2, \ldots, a_m)$, $B = (b_1, b_2, \ldots, b_n)$, of respective lengths m and n, a matrix of m by n elements is computed using the recursion relation:

$$H_{ij} = \max \begin{cases} 0 \\ H_{i-1,j-1} + s(a_i, b_j) \\ H_{i-1,j} + w \\ H_{i,j-1} + w \end{cases}$$

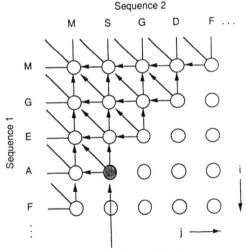

Element (i,j) is dependent on the values
of elements (i-1,j-1), (i,j-1) and (i-1,j)

FIGURE 1   Calculation of the scoring matrix.

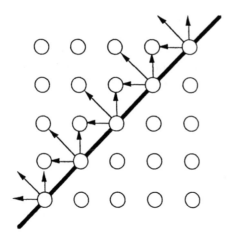

FIGURE 2 All elements on the same antidiagonal can be computed in parallel by a linear array of processors.

where w is the cost of a deletion of one amino acid in one sequence and $s(a_i, b_j)$ is a value for the similarity between amino acids a_i and b_j. The score of the best common subsequence is given by the maximum of $H_{i,j}$ over the matrix and the position of one end in the two sequences is given by i and j. The starting positions of the subsequence can be determined by traceback, which also generates the alignment, or by carrying forward potential start positions during the computation of the matrix. We use the latter method in our implementation.

Note that the computation of element (i, j) depends only on elements $(i, j-1)$, $(i-1, j)$ and $(i-1, j-1)$ so that all elements on the same antidiagonal, where all $(i+j)$ are the same, may be computed simultaneously (Figures 1 and 2). This forms the basis of our parallel implementation in which we configure the CM-2 as a one-dimensional array of processors that computes all $H_{i,j}$ on one antidiagonal simultaneously in one iteration.

One can consider the course of the computation as the traversal of the scoring matrix by the array of processors (Figure 3). The array, lying on an antidiagonal, moves one horizontal step to the right at each iteration and computes all the values on one antidiagonal at each step. This projection of the matrix onto the processor array means that during the course of the computation all elements on a given row of the matrix are computed by a specific processor.

In its simplest form the algorithm requires that every $H_{i,j}$ have five pieces of information supplied to it. These are the identity of the two residues a_i and b_j, the values of $H_{i,j-1}$ and $H_{i-1,j}$ for the horizontal and vertical deletion paths, and the value of $H_{i-1,j-1}$ for the diagonal path. With our mapping of the matrix onto the CM-2, each processor retains the identity of residue a_i throughout the comparison and receives the identity of residue b_j from processor $i-1$ at each step through

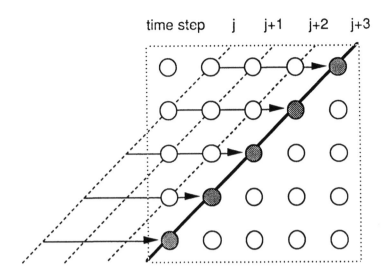

FIGURE 3   On each iteration, the array of processors computes an entire antidiagonal; the matrix is traversed from left to right in successive iterations.

the matrix. Likewise the value of $H_{i,j-1}$, the horizontal deletion path, is simply the value $H_{i,j}$ from the previous step and involves no interprocessor communication. The vertical deletion path requires the value of $H_{i,j}$ from processor $i-1$ on the previous step to be passed to processor i. The value of $H_{i-1,j-1}$ represents the value $H_{i,j}$ of processor $i-1$ from two steps previous. The value is stored within processor $i-1$ for one step and then transferred to processor i during the next step. These steps and the flow of information involved are shown in Figure 4. It can be seen that all the data flows from one end of the processor array to the other, as in a pipe. In order to evaluate the recursion relation described above, each processor must determine a value for the similarity between amino acids a_i and b_j. We store the full similarity matrix in a special form such that one copy is shared between 32 processors. All processors can simultaneously access the appropriate value using indirect addressing into the matrix.

In addition to the flow of information described, we carry forward the position at which $H_{i,j}$ became positive, indicating a nascent matching region, and this can be used to determine the start of the best subsequence at the end of the comparison. Each processor retains the maximum score encountered in the current comparison and the position at which it occurred. After completing one traversal, the maximum across the processors is determined by a single instruction to identify the score and location of the best common subsequence and these are returned to the front end.

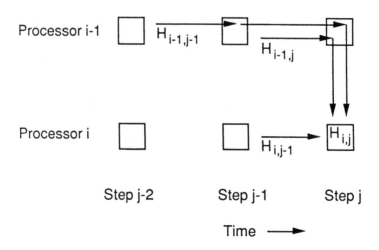

FIGURE 4   The flow of information between processors necessary to compute H_{ij}.

In this manner the entire matrix can be computed in $m + n + 1$ steps. The number of processors in a CM-2 greatly exceeds the average number of residues in proteins and so we are able to perform many comparisons simultaneously. A portion of the database is loaded onto the machine and a single sequence compared to all these sequences in parallel. In this manner data parallelism is used at two levels in the task.

PERFORMANCE

Our current implementation of the algorithm is written in LISP with calls to PARIS and with the inner loop written in microcode. It runs on a Symbolics LISP Machine front-end. We are using this version of the program to perform comparisons between all pairs of proteins in the NBRF database and each comparison returns the score and start and end positions of the best subsequence for every comparison. The CM-2 has an 7 MHz internal clock and the computation of a single matrix entry takes about 700μsec. On a 64K CM-2 we can achieve a theoretical peak performance of 100 Million matrix entries per second and in practice attain 75–80 Million matrix entries per second. By way of comparison one of us (RJ) has written a similar program, but which returns only the score of the best subsequence, for the 3 mips Sun 3/280 computer and has achieved 139,000 matrix entries per second. Jacob Maizel and co-workers have achieved speeds that we estimate correspond to a peak of around 1.1 Million matrix entries per second on a single processor of a CRAY X-MP.[9] The dramatic speed increase achieved through use of a parallel machine is clearly evident from both our figures and those given by Collins and Reddaway in their contribution to this volume. These programs were all written for slightly different purposes and to varying degrees of code optimization, and comparison of the timings should bear these factors in mind. It is clear, however, that our implementation represents a major increase in speed over serial computers and we believe that it has the potential to meet our goal of allowing rapid and sensitive searches with the sequence databases for some time to come, as they rapidly grow in size.

RESULTS

We are currently using the program to perform massive sets of sequence comparisons involving all pairs of proteins in the current protein sequence databases, representing around 50 million comparisons. Such a massive dataset of comparisons provides the means for studying the effectiveness of different metrics for describing amino acid similarity and for studying the statistics of sequence matching, with the aim of being able to directly identify significant relationships from among the majority of unrelated proteins.

Waterman and colleagues[14] have derived probability distributions for the scores expected from comparisons of unrelated nucleic acid sequences. They have shown that for some values of mismatch and gap penalties, the expected score is a linear function of sequence length; for others this function is logarithmic, with a sharp phase transition between the two regions of parameter space. This work has great importance in that it permits the direct assignment of statistical significance to any given DNA sequence comparison. Extending this work to protein sequence comparison would be of even greater value; however, the use of matrices of amino acid similarity in place of simple match or mismatch scores precludes the derivation of the equivalent distributions by the analytic approach and requires empirical studies. Using our sequence comparison code, we have performed all pairwise comparisons of 200 unrelated proteins using a variety of parameter values and have demonstrated the existence of the phase transition in protein sequence comparison. Further study should lead to better understanding of how the results of a comparison are influenced by the various parameters and should permit the direct assignment of statistical significance to any given comparison.

Our major research goal at present is to use these massive sets of comparisons to identify clusters of related proteins and to infer relationships between conserved sequences and functions. Many such families are already known and the NBRF PIR database, for instance, is supplied to users with many of its sequences ordered by family. By our use of sensitive comparison methods, we may be able to identify weaker relationships that have previously been overlooked. In particular, the approach of deriving a multiple sequence alignment from a family of related proteins and using this directly in sequence comparison, as opposed to individual sequences, has been shown to be extremely powerful in identifying weak but biologically significant relationships.[10,12] From such multiple alignments one may also derive consensus sequence templates that may be assigned to particular functions, such as ATP-binding. Collections of such templates have great utility in assigning function to uncharacterized proteins.[13]

We have recently extended the complexity of our algorithm to include affine gap penalties, that is, the ability to penalize the opening of a gap differently than the continuation of an open gap. Insertions and deletions in the sequence alignments of related proteins whose structure is known are often found to occur in loop regions and tend to be excluded from elements of secondary structure, such as helices.[1] This 'concentration' of deletions to certain regions can best be approximated in sequence comparison through the use of affine gap penalties. This involves a fractional increase in the amount of interprocessor communication on the CM-2 and we are currently examining its effect on both performance and the quality of results obtained.

More substantial variants to the code are being considered including the implementation of an algorithm by Gotoh[3] which identifies the best N subsequences in a comparison, as opposed to the single best. This is of great importance in the consideration of weakly related sequences. The method of choice for this purpose would be that of Waterman and Eggert,[15] but as this requires tracebacks and recalculation of the matrix, it is unsuited to a parallel machine. Gotoh's method approximates this algorithm but requires only a forward pass through the matrix, permitting a parallel implementation. The degree to which this method approximates the solution is under current study.

Related to this work we are also developing methods for multiple sequence alignment, consensus pattern identification, comparison of protein structures, and the prediction of structure from sequence.

ACKNOWLEDGEMENTS

Eric Lander was supported in part by National Science Foundation grant #NSF-DCB-8611317 and System Development Foundation grant #SDF612.

REFERENCES

1. Chothia, C., and A. M. Lesk. "The Relation between the Divergence of Sequence and Structure in Proteins." *EMBO J.* **5** (1986):823–826.
2. *Connection Machine Model CM-2 Technical Summary.* Cambridge, MA: Thinking Machines Corporation, 1987.
3. Gotoh, O. "Pattern Matching of Biological Sequences with Limited Storage." *Comp. Appl. Bio. Sci.* **3** (1987):17–20.
4. Hillis, W.D. *The Connection Machine.* Cambridge, MA: MIT Press, 1985
5. Hunkapiller, T., M. S. Waterman, R. Jones, M. Eggert, J. Peterson, and E. Chow. Unpublished work.
6. Lander, E., J. P. Mesirov, and W. Taylor. "Protein Sequence Comparison on a Data Parallel Computer." In *Proceedings of the 1988 International Conference on Parallel Processing.* Philadelphia, PA: Penn State Press, 1988.
7. Lander, E., J. P. Mesirov, and W. Taylor. "Study of Protein Sequence Comparison Metrics on the Connection Machine CM-2." In *Proceedings of Supercomputing '88.* Washington, D.C.: IEEE Computer Society Press, 1989.
8. Lipton, R. J., and D. Lopresti. "A Systolic Array for Rapid String Comparison." In *Chapel Hill Conference on Very Large Scale Integration.* Edited by H. Fuchs. Computer Science Press, 1985.
9. Maizel, Jacob. Personal Communication.
10. Pearl, L. H., and W. R. Taylor. "A Structural Model for the Retroviral Proteases." *Nature* **329** (1987):351–354.
11. Smith, T. F., and M. S. Waterman. "Identification of Common Molecular Subsequences." *J. Mol. Biol.* **147** (1981):195–197.
12. Taylor, W. R. "A Flexible Method to Align Large Numbers of Biological Sequences." *J. Mol. Evol.* **28** (1988):161–169.
13. Thornton, J. M., and W. R. Taylor. "Protein Structure Prediction." In *Peptide and Protein Sequence Analysis.* Edited by M. Geisow and J. Walker. Oxford: IRL Press, in press.
14. Waterman, M. S., L. Gordon, . and R. Arratia "Phase Transitions in Sequence Matches and Nucleic Acid Structure." *Proc. Natl. Acad. Sci. USA* **84** (1987):1239–1243.
15. Waterman, M. S., and M. Eggert. "A New Algorithm for Best Subsequence Alignments with Application to tRNA-rRNA Comparisons." *J. Mol. Biol.* **197** (1987):723–728.

David C. Torney, Christian Burks, Daniel Davison, and Karl M. Sirotkin
Theoretical Biology and Biophysics, Theoretical Division, Group T-10, Mail Stop K710, Los Alamos National Laboratory, Los Alamos, New Mexico 87545

Computation of d^2:
A Measure of Sequence Dissimilarity

An elementary measure of sequence dissimilarity, d^2, is described. The computer algorithm used for its evaluation is discussed in detail. The potential sensitivity of the measure is demonstrated by comparison of sequences with randomly changed letters. The biologic efficacy of d^2 is demonstrated by using the measure to detect the members of the *Alu* repetitive sequence family in sequenced primate DNA, and also to detect the members of the *IS1* repetitive sequence family in sequenced bacterial DNA. We explore a natural weighting scheme for each word's contribution to d^2 and show its utility for finding *Alu* sequences.

INTRODUCTION

The determination of similarity between two biological sequences is of considerable importance; this manuscript discusses a new measure of dissimilarity referred to as d^2. Some mathematical aspects of this measure have been presented.[12] d^2 is derived from the degree to which two sequences share subsequences. A subsequence of n adjacent letters is referred to as an n-word, and all overlapping n-words are considered.

The measure $d_n^2(\underline{\rho})$ is defined by the equation

$$d_n^2(\underline{\rho}) = \sum_{i=1}^{4^n} \rho_n(w_i)\big(m_1(w_i) - m_2(w_i)\big)^2, \tag{1}$$

where w_i is n-word i, $m_j(w_i)$ is the multiplicity of w_i in sequence j, sequences 1 and 2 are compared, and $\rho_n(w_i)$ is a *weight* for n-word i. So long as the weights are non-negative, $\sqrt{d_n^2(\underline{\rho})}$ is a pseudometric measure of sequence dissimilarity for all possible sequences.[3] We will, however, find it useful to allow real $\rho_n(w_i)$, as will be seen below. The summation in Eq. (1) extends over all possible n-words composed of four letters because we will compare DNA sequences, but $d_n^2(\underline{\rho})$ can be computed for an alphabet with an arbitrary number of letters. It may prove useful to combine the values of $d_n^2(\underline{\rho})$ for different values of n. Explicitly,

$$d^2 \equiv \sum_{n=l}^{u} d_n^2(\underline{\rho}), \tag{2}$$

l being the smallest and u being the largest word size contributing to d^2. Our computer algorithm evaluates Eq. (2) with $u \leq 31$ and allows the comparison between a given sequence and a database containing n bases in $4n$ operations, each operation requiring 8.5×10^{-9} seconds on a CRAY X-MP. The prerequisites are described in detail in the next section.

The measure d^2 is similar to a measure referred to as "n-gram," used for document retrieval.[5] In biological sequence analysis, a measure based on 3-word multiplicities has been used to distinguish coding from noncoding sequences.[7] Furthermore, a technique based on the spatial correlations between 1-words has been developed.[9] These indicate that d^2 is a natural measure of sequence dissimilarity and that it is applicable to some biological problems.

COMPUTER ALGORITHM

One starts with *query* sequences to be compared with a database of sequences. The object of the comparison is to determine subsequences of database sequences related to the query. The database sequences are cut into segments w bases long, called *windows*. These windows can be overlapping or non-overlapping as desired; for pedagogy, we take them to be non-overlapping in this section. Eq. (2) is computed for each query sequence and window sequence.

For reasons given immediately below, we expand Eq. (2) to give three terms,

$$d^2 = \sum_{n=l}^{u} \sum_{i=1}^{4^n} \rho_n(w_i) \{ m_D^2(w_i) + m_Q^2(w_i) - 2m_D(w_i)m_Q(w_i) \}$$

where D denotes a database window and Q a query sequence. The first term is the weighted sum of squares of word multiplicities in a database window. The second term is the weighted sum of squares of word multiplicities in a query sequence. The third term is proportional to the weighted sum of products of word multiplicities in both sequences. In the following, we describe how each of these terms is computed, but it is to be noted that only the third term is comparison dependent so that the first two can be computed once for a variety of parameters and stored for use in all future comparisons. Another possibility, which we have not fully explored, would be to use statistical results about the distribution of values of d^2 to decide,[12] based on the value of the third term for a given window and query, whether or not to compute the remaining two terms. Moreover, the determination of the weights can be done once for a given set of database windows and the resulting weights used in subsequent comparisons. The algorithm we use to compute the third term is similar to the *trie* search, used to find the longest exact match between two strings.

In the FORTRAN program GR, nucleotide bases are represented as follows:

$$A = 0, \ C = 1, \ G = 2, \ \text{and } T = 3.$$

All undetermined bases are, for the present, randomly chosen equal to 0, 1, 2, or 3, with probability 0.25. Call the window length w, the query length q, and the smallest and largest j-word lengths used, l and u.

WEIGHTED SUM OF THE MULTIPLICITIES SQUARED FOR WINDOWS

To obtain the multiplicities, create a list of the $(w+1-n)$ n-words, $n = \min(u, w)$, beginning with each base position in the window and going to the right. This list is sorted in ascending numerical order.

Two related lists are used subsequently. One begins with the sorted list of n-words and an ascending list of numbers from 1 to $w + 1 - n$. The lists are now jointly modified by a procedure referred to as *compression*. If two adjacent numbers in the first list are equal, neither the top one nor the corresponding number in the second list is kept. At the end of this process the first list contains sorted n-words with each occurring type represented once, and the second contains the cumulative multiplicities, i.e., the number of n-words whose numerical value does not exceed the corresponding n-word. Thus, the first entry and the differences between subsequent entries in the second list are the n-word multiplicities.

Similar steps for $(n - 1)$-words down to l-words suffice to determine all the necessary multiplicities. The sorted list of next-larger words is shifted two bits (or one base) to the right. The resulting list remains sorted, but it may now contain degenerate entries. Furthermore, one must insert at its sorted position one word arising from the right end of the window because there is one additional word of this size; and the second list is also adjusted. The compression results in a list of words with each occurring type represented only once in the list of cumulative multiplicities. At each wordsize, one either retrieves or computes the word weights. For each window, the weighted sum of the multiplicities squared is stored, to be used when the weighted sum of products of multiplicities is computed.

DATA STRUCTURES FOR EACH QUERY SEQUENCE

The same procedure used to determine the multiplicities of words in windows is repeated for each query sequence. However, in addition to the computation of the query's weighted sum of word multiplicities squared, one creates the IA and AS arrays containing information used to efficiently compute the third term contributing to d^2, involving the products of word multiplicities for all windows compared with a given query. Figure 1 shows the data structures for one query sequence and indicates their relationships. The technique discussed above involving paired lists and compression is used for k-words with $1 \leq k \leq m$, with $m = \min(u, q)$. Concurrently, the words of each length are stored in blocks in the array IV and the products of the weight with the multiplicity at corresponding entries in the array AS. (The weights are the $\rho_k(w_i)$, where w_i is the k-word whose multiplicity is determined.) These

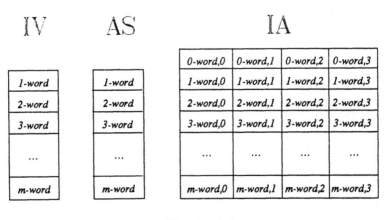

FIGURE 1   Array relations. This is a schematic of the related structures of the three arrays: IV, AS, and IA. The blocks related to n-words are labeled and separated by horizontal dashed lines. The IA array has four entries for each in IV and AS, plus an additional four corresponding to an imaginary 0-mer.

blocks are separated by dashed lines in Figure 1. Within each block the list of words in IV is sorted in ascending order and has each type occurring represented once. For each query, the weighted sum of the multiplicities squared is stored elsewhere, to be used when the weighted sum of products of multiplicities is computed.

Next, the entries in AS corresponding to j-words shorter than l bases are set equal to zero. The array AS is now modified so it contains the product of the weight with the multiplicity of the corresponding j-word in the IV array, plus the sum of products of weight with the multiplicity of all shorter k-words, $l \leq k \leq j$, in the IV array, identical with the rightmost k bases of the j-word. This modification is carried out sequentially, beginning with the comparison between the l-words in IV and the $(l+1)$-words in IV with the leftmost base removed. The latter list is seen to consist of four consecutive sorted sublists, one for each of the bases. The elements of the sublists must occur in the list of l-words. Each of these four sublists is compared separately against the list of l-words, taking advantage of the fact that they are both sorted. One pointer is moved through the sublist and another through the l-word list, beginning at the first entries. If equivalence between the two entries pointed to is found, the entry in the AS array containing the weighted multiplicity of the corresponding $(l+1)$-word has the weighted multiplicity of the l-word pointed to added to it, and the pointer advances one in the sublist. If there is not equivalence, the pointer advances one in the l-word list. This procedure is repeated sequentially with a comparison between $(l+1)$-words and $(l+2)$-words and so on through $(m-1)$-words and m-words. By induction, the desired result, described in the second sentence in this paragraph, is achieved.

The IA array has four entries associated with each word in the IV array and has four entries associated with an imaginary 0-word. The IA entries are the locations of specific words in the IV array. Each of the four entries in IA associated with a k-word in IV results from adding the base 0, 1, 2, or 3 to the right end of the k-word, making a $(k+1)$-word and determining the location of the longest j-word $(j \leq k+1)$, taking the right j bases of the $(k+1)$-word, occurring in the IV array. The location of the imaginary 0-word is taken to be immediately preceding the location of the first 1-word in IV. The four entries associated with the 0-word are the locations in IV of the corresponding bases, for bases occurring in the query sequence. For bases not occurring in the query sequence, the 0-word entries are the location of the imaginary 0-word. To determine the entries in IA corresponding to k-words, with $1 \leq k \leq m-1$, one uses the following procedure, beginning with the 1-words and continuing sequentially to the $(m-1)$-words.

One first asks if the k-word with each of the four bases added to its right end is in the list of $(k+1)$-words. To accomplish this, an expanded list is generated from the k-words in IV with the first four entries being the first k-word catenated with a zero, one, two, or three, and so on. Created in this way, the expanded list remains sorted and nondegenerate. Every $(k+1)$-word in the IV array will occur once in this expanded list. One now loops through the expanded list and uses a pointer in the list of $(k+1)$-words in IV. The pointer begins at the first $(k+1)$-word. For each element in the expanded list, one asks if the $(k+1)$-word pointed to is equal. If so, the location of the $(k+1)$-word in IV is entered at the corresponding entry for the catenated k-word in IA, and the pointer is advanced by one. If not, one reads the rightmost k bases from the element of the expanded list. Since the IA array is complete for 0-words through $(k-1)$-words, one can use it to *read* these k bases, left to right, beginning with the entries in the IA array for the imaginary 0-word. That is, one uses the first base to obtain a location from the corresponding column in the row in IA for the 0-word. One uses the next base to determine the column at the given location in IA from which the next location is read, etc. The result of reading a sequence through the IA array is always that the last location read is the location in the IV array of the longest j-word equal to the last j bases read, precisely what is needed in this situation. The location is the entry for the catenated k-word in IA. The entries in IA for m-words in IV are created according to this latter scheme; that is, they are the locations of j-words not exceeding m in length. An upper bound to the number of entries in the arrays IV, AS, and IA is $6uq$. AS contains up to uq floating-point numbers, and IV and IA contain up to $5uq$ integers. Once the AS and IA arrays are complete, the IV array has no further use.

USING THE DATA STRUCTURES TO COMPUTE THE WEIGHTED PRODUCTS OF MULTIPLICITIES AND SUMMARY OF COMPUTATION

The weighted sum of the multiplicities squared for both the windows and the queries have been determined, and the IA and AS arrays are now used to compute the weighted products of multiplicities. For each base read, based on the current location, a new location is obtained from the IA array, and the value of the AS array at that location is added to a running total. Since the AS array contains the weighted sum of the multiplicities of all j-words in the query sequences matching the right j bases of the indicated k-word, with $l \leq j \leq k$, and the value of k is the longest match with a word in the query sequence ($k \leq m$), the running total at the end of the window is the weighted product of the multiplicities.

We now give a simple example of the use of the data structures to compute the sum of weighted products of multiplicities. The query is the three "letter" sequence 003. We use the following arbitrary weights for the n-words in the query:

$$\rho_1(0) = \rho_1(3) = 0.0; \quad \rho_2(00) = \rho_2(03) = 1.0; \quad \rho_3(003) = 0.25$$

It follows from the above discussion that the elements of the data structures are:

IV	AS	IA
–	0.0	1002
0	0.0	3004
3	0.0	1002
00	1.0	3005
03	1.0	1002
003	1.25	1002

The IV array has elements indexed by 1 through 5; the AS array has elements indexed by 0 through 5; and the IA array has elements indexed by 0 through 3 in each row, and by 0 through 5 in each column. We now use the array IA to read the "target" sequence of six "letters" 100303 and the array AS to determine the sum of weighted products of multiplicities. Locations are computed sequentially, one for each "target" sequence letter.

$$location0 = 0$$
$$location1 = 0 = IA(location0, \text{ target letter } 1)$$
$$location2 = 1 = IA(location1, \text{ target letter } 2)$$
$$location3 = 3 = IA(location2, \text{ target letter } 3)$$
$$location4 = 5 = IA(location3, \text{ target letter } 4)$$
$$location5 = 1 = IA(location4, \text{ target letter } 5)$$
$$location6 = 4 = IA(location5, \text{ target letter } 6)$$

These locations are used to determine the sum of weighted product of multiplicities, which equals

$$\sum_{j=1}^{6} AS(\text{location} j) = 3.25\,,$$

as can easily be determined independently.

Thus, four operations per letter of database sequence are required in this central part of the computation: reading the base, reading one entry of IA, reading one entry of AS, and floating point addition for the running total. These operations are vectorized and each requires one clock period, 8.5×10^{-9} seconds, per letter of database sequence, on the CRAY X-MP. Furthermore, the program GR can compute the weighted products of the multiplicities for multiple query sequences in under two clock periods per letter of database sequence per query, due to "chaining" and distribution. These results are independent of l and u in Eq. (2) and of the weights. The algorithms given in the preceding subsection are reasonably efficient, and establishing the data structures for a query can be accomplished in a small fraction of the c.p.u. time required to scan a megabase-sized database. However, computing the weighted sum of the multiplicities squared for the windows, using the described algorithm, requires orders of magnitude more c.p.u. than all other operations. 0.1 c.p.u. second is required on an X-MP to compute the terms involving products of the multiplicities for a query and three-megabase database. About 10 c.p.u. seconds are required to compute the weighted sum of multiplicities squared for the windows of this three-megabase database. As mentioned earlier, the simplest way to make the c.p.u. requirements of the program GR equal the c.p.u. requirements of computing the sum of the weighted products of the multiplicities is to compute separately the weighted sum of multiplicities squared for the windows. The precomputed values could be used for all subsequent comparisons. Parenthetically, we have not ruled out the possibility of a more efficient algorithm for computing the sum of weighted squares of multiplicities of window words.

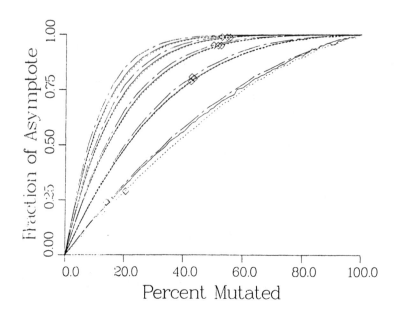

FIGURE 2   Mutation. x-axis: Percent Mutated; $0 \leq x \leq 100$; y-axis: Fraction of Asymptote; $0 \leq y \leq 1.0$. This plot shows the average $d_n^2(\underline{1})$ for a sequence and a mutated version of itself. A *random* sequence (bases chosen independently with probability 0.25 to be each type) is compared with 1000 mutated versions of itself. That is, at every mutation percentage, the corresponding number of distinct base positions in the selected sequence are randomly selected and the corresponding bases are randomly replaced (0.25 probability to be each type), generating one mutated version; and the process is repeated 1000 times. Furthermore, the average is taken of this process repeated ten times with ten different random sequences. Thus, the average results from 10^4 trials. The curves are the result of computing $d_n^2(\underline{1})$ for sequences and mutated versions. The doubly dashed curves are for sequences of length 25; the solid curves are for sequences of length 100; and the dashed curves are for sequences of length 400. The curves of each type with the smallest slope at the origin have n, the word length, equal 2. Similarly, the curves of each type with larger slopes at the origin correspond to $n = 4, 6, 8$, and 10. That n has a larger effect than the sequence length is seen in the five apparent groups of three curves. The resulting averages are *normalized*—divided by the *asymptote* average at 100% mutation. The five asymptotes for the 25 base sequences are 40.5, 44.3, 39.9, 36.0., and 32.0, respectively, for $n = 2$ through $n = 10$. The five asymptotes for the 100 base sequences are 174., 191., 189., 186., and 182., respectively. The five asymptotes for the 400 base sequences are 800., 790., 785., 786., and 782., respectively. In addition, each of these curves of $d_n^2(\underline{1})$ for $n = 2$ through $n = 8$ is marked with a hollow diamond at the y coordinate 1. $- 2 \times$ (average normalized *standard deviation*), the standard deviation being the square root of the variance of distribution of values of $d_n^2(\underline{1})$ at 100% mutation, the normalization being the same as for the curve under consideration, and the average being taken over the 10 *random* initial sequences.

Although we have emphasized performance on a mainframe, the algorithms could be implemented on personal computers as well.

RESULTS AND DISCUSSION

The first example addresses the relations between the amount of "mutation" and the expected value of $d_n^2(\underline{1})$ for two sequences with lengths 25, 100, and 400, and for n equal 2, 4, 6, 8, and 10. Here, all weights equal unity. Figure 2 shows the result, with the *percent mutated* indicating the fraction of the bases (at random positions) that become, with probability 0.25, A, C, G, or T in one of the two initially identical sequences. The initial sequence is constructed by choosing each letter independently and randomly with probability 0.25. The doubly dashed, solid, and dashed curves correspond respectively to sequences of length 25, 100, and 400 bases. The lowest group of three curves shows the average $d_2^2(\underline{1})$; the next group shows the average $d_4^2(\underline{1})$, and so on through $d_{10}^2(\underline{1})$. Details of the averaging procedure and results are given in the figure legend. On the curves for n equal 2 through 8, an open diamond is placed at an ordinate corresponding to two standard deviations down from 1.0 of the distribution of values of $d_n^2(\underline{1})$ for two randomly selected sequences (100% mutation) of the corresponding length.

The sequence length is seen to have a smaller influence than the value of n on the fraction of the expectation of $d_n^2(\underline{1})$ at 100% mutation. Nevertheless, the longer sequences have narrower distributions. For n equal 6 and 8, the diamonds are at abscissas greater than 50% mutation. We will see how $d_6^2(\underline{1})$ performs in the detection of biological sequences, bacterial IS1 sequences sharing 55% of their bases with the query sequences. This would correspond to the 60% mutation in the figure if the mutation density were uniform. Although the idealized mutation experiment and the results shown in Figure 2 differ from what is expected for biological sequences, we can infer that choosing $n = 6$ allows the detection of highly diverged sequences. Larger values of n are increasingly sensitive to fluctuations in the mutation distribution.

$d_6^2(\underline{1})$ is used to detect bacterial *insertion sequences*, IS1, in the three-megabase bacterial subdivision of GenBank DNA sequences.[2] The IS1 sequence and its family members, ISO-IS1's, are found in *E. coli* and other enteric bacteria.[11] They are generically 768 bases in length. Figures 3a, 3b, and 3c show the histograms of the values of $d_6^2(\underline{1})$ when the IS1-R sequence, the IS1-F sequence, and the IS1$v\xi$ sequence are used as the query sequence, respectively. Again, all weights equal unity.

There are five subsequences in the database sharing 99% of their bases with IS1-R; whereas, IS1-F shares approximately 90% of its bases with IS1-R, and IS1-$v\xi$ shares 55% of its bases with IS1-R.[11] The choice of the word length $n = 6$ is based on results shown in Figure 2. The database sequences are cut into overlapping windows 820 bases long with a 205-base interval between windows. This choice of

a

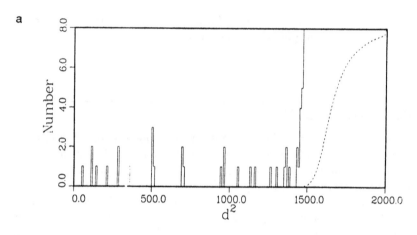

b

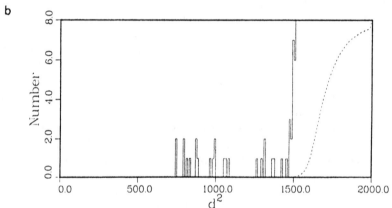

c

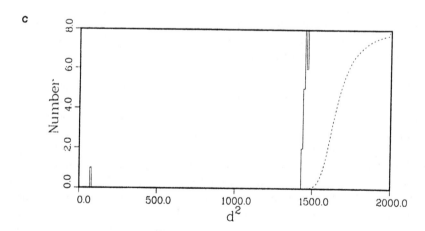

FIGURE 3

FIGURE 3 (continued) IS1 queries, bacterial sequence database. x-axes: $d_6^2(\underline{1})$, $0 \le x \le 2000$; y-axes: Number of windows with $[d_6^2(\underline{1})/10.]$, $0 \le y \le 8$. In each case, a different query is compared against the same windows of 3.2 megabases of sequenced bacterial DNA in GenBank. These windows are 820 bases long, and they are spaced by 205 bases—windows overlap 75%. If necessary, random bases chosen with probability 0.25 are added to complete the last window of each database sequence. In Figures 3a, 3b, and 3c, the query sequences are IS1-R, IS1-F, and IS1-$\nu\xi$, respectively. The histograms of the resulting values of $d_6^2(\underline{1})$ for each query compared with all windows are plotted as the rectilinear curves. The height of the histogram at a given value of $d_6^2(\underline{1})$ equals the number of windows with the integer part of their value of $(d_6^2(\underline{1})/10.)$ equal to the integer part of the given $(d_6^2(\underline{1})/10.)$. In other words, the *bins* of the histogram are ten units wide. The rightmost *bins* of the histogram have for their height the number of windows with $d_6^2(\underline{1})$ not less that 1990. In each figure, the dashed curve is the fraction of windows with smaller values of $d_6^2(\underline{1})$. These curves have been normalized to go between zero and unity; the y coordinate must be divided by eight to give this fraction.

windows ensures maximal sensitivity for d_n^2 because the two compared sequences are approximately the same length, and furthermore, the use of overlapping windows allows one to make sure that windows are approximately aligned with complete IS1 sequences in database sequences. Whereas these windows are clearly discriminated from the large peak as seen in Figures 3a and 3b, they are not in the comparison shown in Figure 3c. Nevertheless, the 20 sequences containing windows with the smallest values of d_6^2 include the five sequences known to be 45% dissimilar. Thus, no known similar sequences are missed in this comparison. More details of using d^2 to find IS1 sequences are given elsewhere.[12]

We now develop a weighting scheme used in the next example. The 4^n weights $\rho_n(w_i)$ occurring in $d_n^2(\underline{\rho})$ are chosen to minimize the value of

$$R \equiv \frac{\text{Variance}(d_n^2\,(\underline{\rho}))}{\left\{\text{Expectation}\left(d_n^2(\underline{\rho})\right)\right\}^2}\,, \qquad (3)$$

for a fictitious query that has uniform word composition compared against a (real) windowed database. The idea is to select the weights based on the word usage density in the windowed database to optimally resolve from the origin the distribution of values of $d_n^2(\underline{\rho})$ from dissimilar sequences, so that similar sequences can be detected. In our example, $n = 6$ and the length of the query of interest equals 300 bases, and the fictitious query is taken to have $295/4^6$ as the multiplicity for every 6-word. (One could choose a different fictitious query or a set of fictitious queries for the purpose of determining the weights for a closely related optimization.) Explicitly, R is

$$R = \frac{\sum_{i,j=1}^{4^n}\rho_n(w_i)\rho_n(w_j)\mu_{ij}}{\left\{\sum_{i=1}^{4^n}\rho_n(w_i)\eta_i\right\}^2} - 1\,, \qquad (4)$$

where

$$\mu_{ij} = \frac{1}{D_T Q_T} \sum_{D=1}^{D_T} \sum_{Q=1}^{Q_T} \left(m_D(w_i) - m_Q'(w_i)\right)^2 \left(m_D(w_j) - m_Q'(w_j)\right)^2, \qquad (5)$$

and

$$\eta_i = \frac{1}{D_T Q_T} \sum_{D=1}^{D_T} \sum_{Q=1}^{Q_T} \left(m_D(w_i) - m_Q'(w_i)\right)^2. \qquad (6)$$

In these formulas, D_T is the total number of windows created from the database, and Q_T is the total number of query sequences used. (In our example, Q_T equals unity.) Inside each squared quantity on the right-hand sides of Eqs. (5) and (6) the first number, $m_D(w_k)$, is the multiplicity of n-word k in the database window indexed by D, and the second quantity, $m_Q'(w_k)$, is the multiplicity of n-word k in the query sequence indexed by Q.

Returning to Eq. (4), its simple form suggests that it can be optimized by using the extremal properties of eigenvalues.[4] In fact, the ability to minimize Eq. (4) using standard matrix analysis, finding the optimal set of weights $\rho_n(w_i)$ subject to a constraint,

$$\sum_{i=1}^{4^n} \{\rho_n(w_i)\}^2 = C, \qquad (7)$$

encourages us to allow the weights to be both positive and negative. Although this quashes the pseudometric property of $d_n^2(\rho)$, only a small fraction of optimally chosen sets of weights will differ in sign from the majority sign. Therefore, from a practical point of view, this loss is not the overriding concern.

To minimize Eq. (4) as a function of the 4^n weights subject to the constraint given in Eq. (7), we employ an orthonormal basis of the matrix μ_{ij}:

$$\sum_{j=1}^{4^n} \mu_{ij} \xi_j^{(k)} = \lambda_k \xi_i^{(k)}; \qquad (8a)$$

$$\sum_{j=1}^{4^n} \xi_j^{(k)} \xi_j^{(l)} = \delta_{k-l}. \qquad (8b)$$

The last symbol, δ_m, equals unity if m equals zero, and equals zero otherwise. We now let

$$\rho_n(w_i) = \sum_{j=1}^{4^n} \alpha_j \xi_i^{(j)} \qquad (9)$$

and substitute into Eq. (4). The result is

$$R = \frac{\sum_{j=1}^{4^n} \alpha_j^2 \lambda_j}{\left\{ \sum_{j=1}^{4^n} \alpha_j \chi_j \right\}^2} - 1 \,, \tag{10}$$

where

$$\chi_j = \sum_{l=1}^{4^n} \xi_l^{(j)} \eta_l \,. \tag{11}$$

Since the α_j's are arbitrary, subject to the same constraint as the $\rho_n(w_i)$, and since R is non-negative, it follows that the λ_j cannot be less than zero; and if any λ_j vanishes, so does χ_j. We can set the corresponding α_j's equal zero and proceed to minimize R as a function of the remaining α's. The numerator is now a positive definite quadratic form in the remaining α's. The extrema of ratios of quadratic forms, where one quadratic form is positive definite, are well understood.[8] In particular, this ratio has one global minimum, given the constraint on the α's. (Since the denominator can vanish, there is no finite maximum.) Taking the first partial derivative of R with respect to all the remaining α's and setting the result equal zero, gives

$$\alpha_j = \frac{a \chi_j}{\lambda_j} \tag{12}$$

as necessary for a stationary point of R. The constant a is fixed by the constraint on the α_j, namely,

$$\sum_{j=1}^{4^n} \alpha_j^2 = C \,. \tag{13}$$

Since there is only one stationary point, subject to the constraint, and R is known to have a global minimum, evaluating R at this point must yield its global minimum. To find the χ_j and λ_j determining the optimal α_j, we use the routine SSIEV[6] to find the eigenvalues, λ_j, and an orthonormal basis of eigenfunctions, $\xi_j^{(k)}$, for the real symmetric matrix, μ_{ij}. From Eqs. (9), (12), and (13), the $\rho_n(w_i)$ are determined. It is to be noted that the global minimum of R is independent of the choice of C used in Eq. (13).

An example employing the weighting scheme is provided by taking the 300-base-long *Alu* consensus sequence[1] as the query sequence (deleting a few terminal *A*'s) and the 3.8 megabase subsection of GenBank, containing primate DNA sequences, as the database. The database sequences are cut into non-overlapping windows 100 bases long, and $d_6^2(\rho)$ is computed for each query-window combination. The weights were determined using these windows and $n = 6$. The ratio of lengths 300:100 en-sures that fragmentary *Alus* of approximately 100 bases might be detected and also that some windows will lie within complete *Alus*. The sensitivity of d^2 is expected to decrease as the ratio of lengths increases, but it is reasonable to compute d^2 be-tween database windows and queries that are order of magnitude the same length.

a

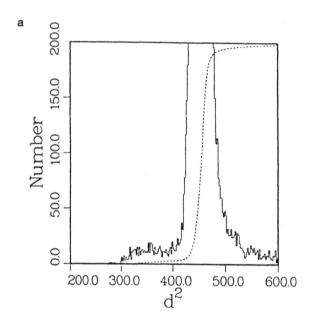

b

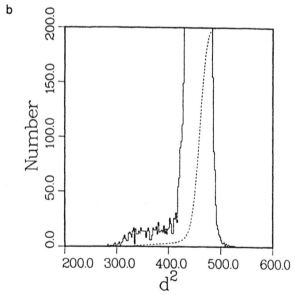

FIGURE 4

FIGURE 4 (continued) Alu query, primate sequence database. x-axes: $d_6^2(\underline{\rho})$, $200 \leq x \leq 600$; y-axes: Number of windows with $[d_6^2(\underline{\rho})/2.]$, $0 \leq y \leq 200$. In both cases, the 300-base Alu consensus sequence is the query sequence compared against windows from 3.8 megabases of sequenced primate DNA in GenBank. These windows are 100 bases long, non-overlapping, and contiguous. If necessary, random bases, chosen with probability 0.25, are added to complete the last window of each database sequence. The values of $d_6^2(\underline{\rho})$ are computed; in Figure 4a the weights $\rho_6(w_i)$ equal unity, and in Figure 4b the weights are the optimal weights, with the normalization coefficient in Eq. (13) chosen so that the two histograms have the same expectations. The histograms of the resulting values of $d_6^2(\underline{\rho})$ are plotted as the rectilinear curves. The height of the histogram at a given value of $d_6^2(\underline{\rho})$ equals the number of windows with the integer part of their value of $(d_6^2(\underline{\rho})/2)$ equal the integer part of the given value of $(d_6^2(\underline{\rho})/2)$. In other words, the $bins$ of the histograms are two units wide. The leftmost $bins$ of the histograms include all windows with $d_6^2(\underline{\rho})$ less than 202; whereas, the rightmost bins of the histograms include all windows with $d_6^2(\underline{\rho})$ not less than 598. In each figure the dashed curve is the fraction of windows with smaller values of $d_6^2(\underline{\rho})$. These curves have been normalized to go between zero and unity; the y coordinate must be divided by 200 to give this fraction.

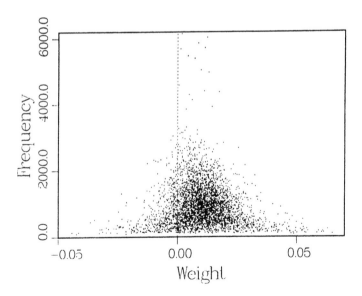

FIGURE 5 Correlation plot of optimal word weight and word frequency. The number of occurrences of each 6-word is determined in the database. The weight for each 6-word follows from Eqs. (9), (12), and (13). The 6-word 000000 is not plotted here; its frequency is approximately 10,500 and its weight is 3.04×10^{-4}.

Figure 4a shows the histogram of resulting values of $d_6^2(\underline{1})$, that is, with all the weights equal unity. These are 805 windows with a value of $d_6^2(\underline{1})$ less than or equal 420. Of these, about 100 windows (with values of $d_6^2(\underline{1})$ principally between 400 and 420) may not contain sequences biologically related to the *Alu* sequence.[12] Figure 4b shows the histogram of resulting values of $d_6^2(\rho)$, where the 4096 $\rho_6(w_i)$ have been chosen in accord with Eqs. (9), (12), and (13) and where the normalization constant C has been chosen so that the expected value of $d_6^2(\rho)$ equals the expected value of $d_6^2(\underline{1})$. The histogram in Figure 4b has a reduction in its variance by a factor of 7 in comparison with the histogram in Figure 4a. This is due largely to the suppression of the contribution from 6-words occurring in *tandem repeats* that are responsible for the tail of the histogram for large values of $d_6^2(\underline{1})$ seen in Figure 4a but not in Figure 4b. Tandem repeats are periodic arrays of bases. Returning to the windows containing *Alus* in Figure 4b, there are 850 windows with values of $d_6^2(\rho)$ less than 421. Of these, 160 windows are not included in the 805 windows seen to be potentially *Alus* in the comparison with uniform weights. However, over half of these 160 windows are adjacent to windows in the remaining 690 windows, an indication of the increased sensitivity afforded by the optimal choice of the $\rho_6(w_i)$. Taken together, it is likely that slightly less than 2% of the windows contain unambiguous similarity to the query *Alu* sequence. Since a similar result is obtained using the reverse complement of the *Alu* consensus sequence as the query (data not shown), this result is in approximate agreement with the fraction of the database and the human genome determined to be *Alu*-like by a variety of experimental and computational approaches.[10] Furthermore, these histograms show no trend toward increasing numbers of increasingly dissimilar *Alu* sequences in the database.

Figure 5 is a correlation plot that puts a point for each 6-word at the coordinate [$\rho_6(w_i)$, number of occurrences of w_i in the database]. The distribution of values of $\rho_6(w_i)$ for words having the same frequency is quite large, indicating the inappropriateness of basing a weighting scheme on the frequencies. The $\rho_6(w_i)$ are responsive to the fluctuations in the density of the frequencies of 6-words in windows. Greater than 86% of the $\rho_6(w_i)$ are positive, and the positive weights are, on the average, almost twice as large.

SUMMARY

This manuscript contains a preliminary development of the measure d^2. We have focused on its computation and also on a possible word weighting scheme. Incidentally, this scheme provides a basis for including words of different lengths in d^2. The examples given demonstrate that d^2 can be successfully applied to biological problems and that the weighting scheme is useful in this context.

ACKNOWLEDGMENTS

We thank Micah Dembo, Walter Goad, Tony Warnock, and Michael Waterman for their comments. We thank Angel Garcia for providing access to the NMFECC CRAY-2, on which the 4^6 optimal 6-word weights were determined. D.C.T., C.B., and D.D. received support from NIH grant #GM-37812. D.D. is a U.S. Department of Energy Alexander Hollaender Distinguished Postdoctoral Fellow. We commend Patricia Reitemeier for manuscript preparation.

REFERENCES

1. Baines, W. "The Multiple Origins of Human *Alu* Sequences." *J. Mol. Evol.* **23** (1986):189–199.
2. Burks, C., and others. "GenBank: Current Status and Future Directions." *Methods in Enzymology*, in press.
3. Copson, E. T. *Metric Spaces.* Cambridge Tracts in Mathematics and Mathematical Physics, Vol. 57. Cambridge, England: University Printing House, 1968.
4. Courant, R. and D. Hilbert. *Methods of Mathematical Physics*, vol. 1. New York: Interscience Publishers, 1953, Chapter 1.
5. D'Amore, R. J. and C. P. Mah. "One-Time Complete Indexing of Text: Theory and Practice." *Proceedings of the Eighth Annual Conference of the S.I.G.I.R. of the ACM (Montreal, Canada).* New York: Association for Computing Machinery, 1985, 155–164.
6. Dongarra, J. J., C. B. Moler, J. R. Bunch, and G. W. Steward. *LINPACK User's Guide.* Philadelphia: SIAM, 1979.
7. Fichant, G. and C. Gautier. "Statistical Method for Predicting Protein Coding Regions in Nucleic Acid Sequences." *CABIOS* **3** (1987):287–295.
8. Gantmacher, F. R. *The Theory of Matrices*, vol. I. New York: Chelsea Publishing, 1959, Chapter X, Sections 6 and 7.
9. Mironov, A. A. and N. N. Alexandrov, "Statistical Method for Rapid Homology Search." *Nucleic Acids Research* **16** (1988):5109–5173.
10. Moyzis, R. K., D. C. Torney, J. Meyne, J. M. Buckingham, J. R. Wu, C. Burks, K. M. Sirotkin, and W. B. Goad. "The Distribution of Interspersed Repetitive DNA Sequences in the Human Genome." *Genomics* **4** (1989):273–289.
11. Ohtsubo, E., H. Ohtsubo, W. Doroszkiewisz, K. Nyman, D. Allen, and D. Davison. "An Evolutionary Analysis of Iso-IS1 Elements from *Escherichia coli* and *Shigella* Strains." *J. Gen. Appl. Microbiol.* **30** (1984):359–376.
12. Torney, D. C., C. Burks, and D. Davison. "A Simple Measure of Sequence Dissimilarity." *SIAM J. Appl. Math.*, submitted for publication.

Michael S. Waterman†‡ and Louis Gordon†
†Department of Mathematics, University of Southern California and ‡Department of Molecular Biology, University of Southern California

Multiple Hypothesis Testing for Sequence Comparisons

INTRODUCTION

In 1970 dynamic programming was first applied to the comparison of biological sequences by Needleman and Wunsch.[8] Their method is now called a similarity method. Since their work, many extensions and modifications have been introduced. This includes distance methods, general gap functions, multiple alignment procedures, and near-optimal methods. See Waterman[14] for a review of these approaches to sequence comparison.

When more sequences began to appear in the later 1970's, it became apparent that alignment of entire sequences was frequently not the major problem of interest. Instead it was more valuable to look for the good matching segments within longer sequences. Distance methods had become very popular, perhaps due to their mathematical relationship to metric spaces, and converting distance methods to handle segmental matching was a difficult task. See Sellers[9] and Goad and Kanehisa.[6] Similarity methods however could more easily be modified for segment comparisons.[10,12] Next we present the algorithm from the papers. Take the two sequences to be $\underline{a} = a_1 a_2 \ldots a_n$ and $\underline{b} = b_1 b_2 \ldots b_m$. They can be either DNA or protein sequences. The similarity measure between sequence letters a and b is $s(a,b)$, where $s(a,b) > 0$ if $a = b$ and $s(a,b) < 0$ for at least some cases of $a \neq b$.

Computers and DNA, SFI Studies in the Sciences of Complexity, vol. VII
Eds. G. Bell and T. Marr   1990   **127**

Insertions or deletions of length k receive weight $-w(k)$. The observation of Smith and Waterman[10] is that negative scoring alignments are of no interest. $S(\underline{a}, \underline{b})$ is defined to be the best (largest) score from aligning $\underline{a}$ and $\underline{b}$. Define

$$G_{i,j} = \max\left\{0; S(a_x a_{x+1}\ldots a_i, b_y b_{y+1}\ldots b_j) : 1 \le x \le i, 1 \le y \le j\right\}. \quad (1)$$

$G_{i,j}$ is the best score of any alignment ending at a_i and b_j or 0, whichever is larger. The similarity algorithm is started with $G_{i,0} = G_{0,j} = 0$ for $1 \le i \le n$ and $1 \le j \le m$. Then

$$G_{i,j} = \max\left\{0, G_{i-1,j-1} + s(a_i, b_j), E_{i,j}, F_{i,j}\right\}, \quad (2)$$

where

$$E_{i,j} = \max_{1 \le k \le j}\left\{G_{i,j-k} - w(k)\right\}, \quad (3)$$

and

$$F_{i,j} = \max_{1 \le k \le i}\left\{G_{i-k,j} - w(k)\right\}. \quad (4)$$

The best scoring alignment has score $\max G_{ij}$. Gotoh[7] showed the time for the multiple gap algorithm of Waterman et al.[11] could be reduced to $O(n^2)$ for linear gap functions $w(k) = u + vk$. For the above algorithm this is accomplished by altering the last two recursions, Eqs. (3) and (4), to:

$$E_{i,j} = \max\left\{G_{i,j-1} - (u+v), E_{i,j-1} - v\right\},$$
$$F_{i,j} = \max\left\{G_{i-1,j} - (u+v), F_{i-1,j} - v\right\}.$$

The average sequence in GenBank or EMBL is 1000 bases long. Figure 1 shows best segment alignments for independent simulations of 10 independent pairs of length 1000 DNA sequences with $P(A) = P(C) = P(G) = P(T) = .25$. The algorithm parameters are $s(a,a) = 1$, $s(a,b) = -\mu$ for $a \ne b$, and $w(k) = \delta k$, where $\mu = 1.1$ and $\delta = 2.1$. It is remarkable that these segmental matchings from random sequences are so long and score so well. Simulations such as this suggest that understanding the distribution of score ($\max G_{ij}$) under the null hypothesis of independence is an important goal. Otherwise if the analysis of "interesting" alignments proceeds on an *ad hoc* basis, it is easy to be misled by statistically insignificant alignments. As the genome projects get underway and megabases of sequence are produced, these statistical considerations will assume more importance. The examples of this paper are of DNA sequences, but the general theory allows analysis of protein and other sequences.

Alignment	max G_{ij}	matches	mismatches	indels
GGCAG-TCTTAGAA ||||| |||||||| GGCAGCTCTTAGAA	10.9	13	0	1
GGCTGATCGAGCGAGGC |||| ||| |||||||| GGCT-ATCTAGCGAGGC	11.4	15	1	1
GATTCAAGGACCAGATAGAGT |||| || | ||||||||||| GATTTCAGAAACAGATAGAGT	11	17	4	0
CAGCGAAAATGCAACGCC ||||| || | ||||||| CAGCGCAACT-CAACGCC	9.9	15	2	1
AGGA-CGAATATCA-GTATAACGATGACG |||| |||||| || | ||||| || ||| AGGATCGAATACCATGGATAAC-AT-ACG	11.6	23	2	4
CGAGCCCTCCGT |||||||||||| CGAGCCCTCCGT	12	12	0	0
CCCGGATGCGCAGGG |!|||||||||| ||| CCCGGATGCGC-GGG	11.9	14	0	1
AACAGCTTATA |:||||||||| AACAGCTTATA	11	11	0	0
AGATTA-TCAATCCA-CGT-GCG |||||| | |||||| ||| ||| AGATTACTGAATCCATCGTAGCG	11.2	19	1	3
TAGTACTCTACTGGTC |! ||||||||||| || TAATACTCTACTGCTC	11	14	2	0

FIGURE 1   Simulation results from comparing 10 pairs of independent, identically distributed DNA sequences of length $n = 1000$ with equally likely letters. The algorithm parameters are $\mu = 1.5$ and $\delta = 2.1$.

PROBABILITY DISTRIBUTIONS

When two random sequences of length n ($\underline{a}$ and $\underline{b}$) are written in a fixed alignment, the resulting sequence of matches and mismatches can be identified as a sequence of coin tosses. The probability that the kth toss is a head is $P(x_i = y_i)$. Our object in this paper is to study long runs of matches, which in this case are long head runs. The celebrated Erdös-Rényi law[5] gave order magnitude behavior for the longest run of heads in a sequence of n independent coin tosses. Their results actually include behavior of the longest head run containing $(1 - \alpha) \times 100\%$ tails where $\alpha > P(H) = p$. For length R_n of pure head runs ($\alpha = 1.0$) their result is

$$\frac{R_n}{\log_{1/p}(n)} \to 1 \text{ with probability one},$$

while for general $\alpha > p$ their result is

$$\frac{R_n}{\frac{\log(n)}{H(\alpha,p)}} \to 1 \text{ with probability one},$$

where $H(\alpha, p) = \alpha \log(\alpha/p) + (1 - \alpha) \log((1 - \alpha)/(1 - p))$ is relative entropy. For $\alpha = 1$, $H(\alpha, p) = \log(1/p)$ and the two results are consistent. Other work[2] gives precise results for this law and gives

$$E\{R_n\} = \log_{1/p}(n) + \frac{(.577\ldots)}{\theta} - \frac{1}{2} + r_1(n) \tag{5}$$

and

$$\mathrm{Var}\{R_n\} = \frac{\pi^2}{6\theta^2} + \frac{1}{12} + r_2(n), \tag{6}$$

where $.577\ldots$ is the Euler-Mascheroni constant, $\theta = \ln(1/p)$, and the remainders $r_1(n)$ and $r_2(n)$ are of very small, but nonvanishing magnitude. For 512 fair coin tosses, the mean is approximately 9.33 while the standard deviation is about 1.93.

The probability results quoted here are for independent and identically distributed coin tosses. To carry the results over to sequence matching, the two sequences of length n are assumed to have bases chosen independently and identically with $p = P\{\text{two bases match}\} = p_A^2 + p_C^2 + p_G^2 + p_T^2$. The formulae (5) and (6) hold with n replaced by n^2.

To consider approximate matching, we allow a fixed number of mismatches k. The mean length of the longest match with k mismatches becomes, from Ref. (2), approximately

$$\log(qn^2) + k \log \log(qn^2) + k \log\left(\frac{q}{p}\right) - \log(k!) + k + \frac{.577\ldots}{\theta} - \frac{1}{2} \tag{7}$$

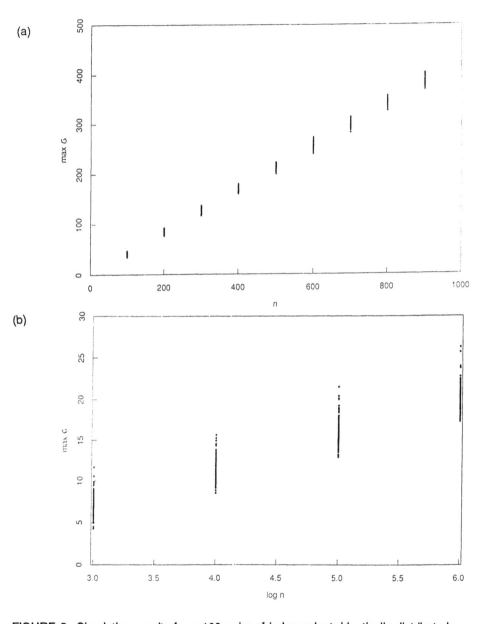

FIGURE 2 Simulation results from 100 pairs of independent, identically distributed DNA sequences of length $n = 1000$ with equally likely letters are compared for (a) $n = 100$ to 900 (growth rate is linear; the algorithm parameters are $\mu = 0.2$ and $\delta = 0.5$) and for (b) $n = 4^3, \ldots, 4^6$ (growth rate is logarithmic; the algorithm parameters are $\mu = .9$ and $\delta = 2.0$).

where $q = 1 - p$ and all log's are taken to base $1/p$. The variance remains approximately $\pi^2/(6\theta^2) + 1/12$.

The Erdös-Rényi law for the length R_n of the longest 100% head run of n coin tosses then extends to a law for the length M_n of the longest match between two sequences. We have recently shown that the length of the longest run of matches containing $(1 - \alpha) \times 100\%$ mismatches satisfies

$$\frac{M_n}{\frac{\log(n^2)}{H(\alpha,p)}} \to 1 \text{ with probability one},$$

which is the Erdös-Rényi law with n replaced by n^2. This last theorem has been obtained by use of the theory of large deviations.[4]

Considering these results, it is not surprising that $\log(n)$ laws hold far beyond the longest exact head run or match. The expected behavior of $\max G_{ij}$ is of importance in evaluating sequence comparisons. If a located match is at or below that expected from random sequences of similar composition, then the match should not be further considered without additional biological information. These distributions have been shown to fit biological sequences quite well[14] for algorithm parameters not covered by the theorems above. We have also proven that $\max G_{ij}$ undergoes a phase transition.[13] For larger values of (μ, δ), $\max G_{ij}$ grows proportional to $\log(n)$ and for smaller values $\max G_{ij}$ grows linearly with sequence length. There are only two modes of behavior at this precision. The logarithmic and linear regions of this two-dimensional parameter space have been determined numerically in a Monte Carlo study.[14] See Figure 2 for illustrations of the growth of score ($\max G_{ij}$) with sequence length. These results help those analyzing macromolecular sequences to proceed in a much less *ad hoc* manner.

A HEURISTIC CALCULATION

Our results provide substantial intuition when confronting more complicated problems as well. As an example, we provide a tentative analysis of a multiple inference procedure presented in Altschul and Erickson.[1] They propose as a measure of sequence similarity the minimal attained significance among all runs of matches, minimizing over all possible run lengths. Attained significance is computed using a binomial model, implicitly computing significance under the model of independently generated letters in each of the compared sequences. Citing a lack of theoretical development, they use a curve-fitting approach to parametrize their proposed test. In the discussion below, we will show how the previously discussed results suggest a framework in which an alternative, theoretically motivated parametrization provides a better fit.

Specifically, in terms of the notation of the previous section, let $M_n(k)$ denote the length of the longest match between two sequences of length n of independently generated letters taken from the same distribution. Inspired by Cramer's treatment of large deviations theory, approximate the log-probability of observing k or fewer mismatches in match length t as

$$-tH\left(1 - \frac{k}{t}, p\right) - \frac{1}{2}\ln(2\pi k),\qquad(8)$$

an approximation good to order k/t. Note that for fixed k the attained significance is increasing in t. Hence, to a crude order of approximation, the logarithm of attained significance might be expected to be

$$S_n = -M_n(K)H(1, p) - \frac{1}{2}\ln(2\pi K),$$

where the random variable K indexes the most significant of all k-interrupted matches between the two n-sequences. Of importance is only the observation that approximate significance will pick one of the longest k-interrupted matches, each of which is approximately distributed as integerized extreme value with centers given by Arratia et al.[2]

There is reason to believe that the most significant of the k-interrupted matches should appear with relatively small k. The argument depends on the approximate independence—for small k—of the lengths of the longest k-interrupted runs of matches. This is a consequence of the conditional uniform distribution of the locations of mismatches given an extremely rich pattern of matches. A common renewal-theoretic clumping argument suggests that for large k, one should expect to observe clumps of unusually long k-interrupted runs of matches. Hence, the most significant among the k-interrupted runs should typically occur for small k, and we use Eq. (7) to conjecture the form

$$S_n = \ln(n^2) + \alpha \ln\ln(n^2) + \beta\frac{\ln\ln(n^2)}{\ln(n^2)} + \gamma + V,\qquad(9)$$

where V is the integerized extreme value with approximate variance $\pi^2/6 + 1/12 = 1.73$, $\alpha = E\{K\}$, and $\gamma = E\{\mu_K\}$ consolidates those terms of Eq. (7) which are not

TABLE 1 Three Models by Ordinary Least Squares

model	$\sqrt{R^2}$	$1000 \times$ MSE
1. $s - \ln(n^2) = \alpha \ln\ln(n^2) + \beta$	.95	3.02
2. $s - \ln(n^2) = \alpha \ln(n^2) + \beta$	.96	2.34
3. $s - \ln(n^2) = \alpha \ln\ln(n^2) + \beta\frac{\ln\ln(n^2)}{\ln(n^2)} + \gamma$	.98	1.44

coefficients of $\ln(t)$ or $\ln\ln(t)$. The correction term with coefficient β is of the sort used in Arratia et al.[2] to correct for finite sample size. For a theoretical discussion showing the form of such corrections, see de Bruijn,[5] section 2.4.

We present the results of fitting Eq. (9) using the data of Altschul and Erickson,[1] in which are reported means of 1000 simulation experiments realizing the minimal attained significance for nine levels of n ranging from 70 to 518. Specifically, we take the sample mean of the 1000 attained significance levels for each value of n using the formula $s = u + .5772/\lambda$, where values of u and λ are taken from Altschul and Erickson,[1] Table II. We fit the three models in Table 1 by ordinary least squares. All three models regress against the independent variable $s - \ln(n^2)$, motivated by the heuristic argument above. It is nearly impossible to select among the models for data analytic reasons alone; correlations among pairs of the three explanatory variables used all exceed .997 in absolute value.

The reassuring feature of the heuristic specification of model 3 is the mean square error after fitting. The tabled values would estimate the variance of an integerized extreme value, if the specification, Eq. (9), were correct. Note that 1.44 is closest of all three specifications to $\pi^2/6 + 1/12$, although all three models yield confidence intervals for the variance which include 1.73.

ACKNOWLEDGMENTS

This work was supported by the National Science Foundation and the National Institutes of Health.

REFERENCES

1. Altschul, S. F., and B. W. Erickson. "A Nonlinear Measure of Subalignment Similarity and Its Significance Levels." *Bull. Math. Biol.* **48** (1986):617–632.
2. Arratia, R., L. Gordon, and M. Waterman. "An Extreme Value Theory for Sequence Matching." *Ann. Statist.* **14** (1986):971–993.
3. Arratia, R. A., L. Gordon, and M. S. Waterman. "The Erdös-Rényi Law in Distribution, for Coin Tossing and Sequence Matching." Submitted to *Ann. Statist.*.
4. Arratia, R., and M. Waterman. "The Erdös-Rényi Strong Law for Pattern Matching with a Given Proportion of Mismatches." *Ann. Prob.*, to appear.
5. de Bruijn, N. G. *Asymptotic Methods in Analysis*. New York: Dover, 1981.
6. Goad, W. B., and M. I. Kanehisa. "Pattern Recognition in Nucleic Acid Sequences. 1. A General Method for Finding Local Homologies and Symmetries." *Nucl. Acids Res.* **10** (1982):247–263.
7. Gotoh, O. "An Improved Algorithm for Matching Biological Sequences." *J. Mol. Biol.* **162** (1982):705–708.
8. Needleman, S. B., and C. D. Wunsch. "A General Method Applicable to the Search for Similarities in the Amino Acid Sequences of Two Proteins." *J. Mol. Biol.* **48** (1970):444–453.
9. Sellers, P. "The Theory and Computation of Evolutionary Distances: Pattern Recognition." *J. Algorithms* **1** (1980):359–373.
10. Smith, T. F., and M. S. Waterman. "Identification of Common Molecular Subsequences." *J. Mol. Biol.* **147** (1981):195–197.
11. Waterman, M. S., T. F. Smith, and W. A. Beyer. "Some Biological Sequence Metrics." *Adv. Appl. Math.* **20** (1976):367–387.
12. Waterman, M. S., and M. Eggert. "A New Algorithm for Best Subsequence Alignments with Application to tRNA-rRNA Comparisons." *J. Mol. Biol.* **197** (1987):723–728.
13. Waterman, M. S., L. Gordon, and R. Arratia. "Phase Transitions in Sequence Matching and Nucleic Acid Structure." *Proc. Nat. Acad. Sci.* **84** (1987):1239–1243.
14. Waterman, M. S. *Mathematical Methods for DNA Sequences*. Boca Raton, FL: CRC Press, 1989.

Analysis

Craig J. Benham
Department of Biomathematical Sciences, Mount Sinai School of Medicine, 1 Gustave Levy Place, New York, NY 10029

Superhelical Transitions and DNA Regulation

SEQUENCE-SPECIFIC DNA POLYMORPHISMS

In addition to carrying the genetic message, DNA has the potential to encode other information. One way this is accomplished that is important for regulation involves sequence-specific susceptibilies to conformational transitions. Although all DNA base pairs prefer to exist in the B-form, certain sequences also can occur in other secondary structures. For example, runs in which the purine and pyrimidine bases alternate along each strand of the duplex can adopt the left-handed Z-form.[27] Sequences having inverted repeat homology can form cruciforms.[15] Homopyrimidine runs of sufficient length can exist in the H-form, which recent evidence suggests is a partially triple-stranded conformation.[19] Although any base sequence can denature,[32] the energy required for this strand separation is smaller for an A·T-rich run than for a G·C-rich one. Coded into each DNA sequence is a collection of specific sites where the molecule can assume particular types of alternative conformations.

The B-form structure is energetically favored in unconstrained molecules under physiologically reasonable environmental conditions. Transitions to alternative conformations must be induced either by changes of environmental parameters (such as ionic strength, temperature or solvent) or by imposition of constraints which act to stress the molecule. In physiological DNA it is the topological constraints imposed on the molecule by the manner of its confinement that regulate imposed stresses, and hence conformational transitions.

DNA within living systems is held in looped (or occasionally closed circular) structures called topological domains. In eukaryotes these domains are approximately 50,000 base pairs long,[33] while in prokaryotes the average domain size is somewhat smaller.[29,30] This subdivision into looped domains usually is effected by periodic attachments of the DNA to a scaffolding in a way that precludes the local rotation of the duplex about its axis at the attachment site. In this case the domains comprise the contiguous segments of DNA extending between attachment sites. This arrangement allows the topological constraints imposed on separate domains to be independently regulated.[1]

Under physiological conditions, *in vivo* domains of DNA commonly are maintained in a negatively superhelical state, in which the linking number is smaller than its relaxed value[1]:

$$\Delta Lk = Lk - Lk_0 < 0.$$

This is accomplished in part through the agency of topoisomerase enzymes, which alter the linking number by mechanisms involving transient strand breaks. The resulting linking deficiency must be accommodated by changes either in the molecular duplex twist, Tw, or in the large-scale tertiary structure as expressed by the writhing number Wr[9,14,35]:

$$\Delta Lk = \Delta Tw + \Delta Wr < 0.$$

Writhing requires bending, while changes of twist involve elastic torsional deformations and/or local transitions to alternative conformations. All of these responses require free energy, so the stressed equilibrium of a superhelical molecule will be determined by the competition among alternative ways of accommodating the imposed linking difference.

In the physiological case of a linking deficiency, transitions to alternative conformations having smaller right-handed helicity than the B-form can occur because they result in a local decrease of the duplex twist, $\Delta Tw_{trans} < 0$. This isolates part of the linking deficiency as a decrease of right-handed twist at the site of transition, which allows the balance of the domain to relax a corresponding amount. Transitions of this type will become favored to occur at equilibrium when the deformation strain energy they relieve exceeds their energetic cost. Because all of the alternate conformations described above are less twisted (in the right-handed sense) than the B-form, transitions to any of these structures can be driven by a negative linking difference. This theoretical prediction has been amply confirmed for each type of transition by *in vitro* experimentation.[19,21,24,26,31,32] More recently, transitions both to Z-form and to cruciforms have been demonstrated to occur *in vivo*.[11,16,18,25]

THE ANALYSIS OF TRANSITIONS IN STRESSED DNA

The equilibrium distribution of a population of identical DNA molecules amongst its available states may be computed theoretically using standard statistical mechanical methods.[13] A partition function is calculated by enumerating states, ascribing a free energy to each, and summing their associated Boltzmann factors:

$$Z = \sum_i \exp(-G_i/kT).$$

Here i indexes the states and G_i denotes the free energy of state i. The probability (i.e., fractional occupancy) of state i in a population of identical molecules at equilibrium is given by the ratio of its Boltzmann factor to the partition function:

$$p_i = \frac{\exp(-G_i/kT)}{Z}.$$

If a parameter γ has value γ_i in state i, then its population-averaged value at equilibrium is:

$$\overline{\gamma} = \sum_i p_i \gamma_i.$$

In this way the equilibrium value of any parameter of interest can be calculated, at least in principle. These may include probabilities of transition of specific base pairs, expected number of transformed pairs, expected number of runs of transition, probability of having any specific number of transformed runs or pairs, expected amount of total twist absorbed by the transition, etc.

Numerous analyses of specific transitions have been performed to date.[2-8] illustrate both the general principles common to all superhelically driven transitions, and the specific details of particular transitions. Because this work has been published extensively elsewhere, we only summarize the most important results here.

First, transitions to any of the above types of alternative secondary structures can be driven by negative superhelicity in molecules containing sites susceptible only to that transition.[2,3,4,6] In domains containing multiple sites susceptible to different transitions, their competition at equilibrium can be complex.[5,7,8] The dominant determinant of transition behavior is the free energy relieved elsewhere by the transconformation reaction. This means transitions to Z-form, which localize about 65 degrees of undertwist per transformed base pair, will occur at less extreme linking deficiencies than alternative transitions, other factors remaining fixed. Because there is a relatively high energetic cost associated with initiation of a new region of transition, at equilibrium there will be only a small number of runs of transition. Specific transitions may occur at equilibrium only in a narrow range of linking differences ΔLk. The reversion of one transformed region back to B-form can be coupled to a conformational transition at some other site in the domain as the linking deficiency is made more extreme.[6,7]

SUPERHELICITY AND REGULATION IN LIVING SYSTEMS

Much of the negative linking difference imposed on DNA domains *in vivo* is stabilized through interactions with basic proteins. In eukaryotes the DNA wraps around histone octomers to form nucleosomes.[30,33] Prokaryotic DNA is similarly constrained by interactions with HU proteins.[12,30] Although the precise nature of this association is not completely understood at present, it is known to be both looser and less extensive in prokaryotes than in eukaryotes. Only about half the DNA is associated with HU proteins in *E. coli*. The constraint on tertiary structure imposed by these interactions stabilizes much of the negative superhelicity of the DNA within domains.

The association of the DNA with proteins in both prokaryotes and eukaryotes permits non-enzymatic methods of modulating superhelical stresses. Changes in the number of associated nucleosomes will alter the amount of ΔLk that is restrained as tertiary structural windings. Also, conformational transitions within or relative motions of the nucleosomal particles may modulate stresses on regions of the DNA. Theoretical analysis suggests that superhelically driven transitions are expected to occur at less extreme superhelicities in DNA whose tertiary structure is constrained than in unconstrained molecules.[10]

In prokaryotes not all superhelicity is stabilized by association with HU proteins. In these organisms the linking difference is known to be modulated in such a way that stresses on the DNA result. At any one time approximately half the domains in *E. coli* experience torsional stress which can drive transitions.[30]

In eukaryotes the situation is less clear. It appears that the majority of the DNA within the eukaryotic genome is not superhelically stressed. However, there is compelling evidence that some DNA domains are stressed *in vivo*. In particular, domains containing actively transcribing genes are stressed.[22] Superhelical stresses are not the norm in eukaryotic DNA largely because any particular type of differentiated eukaryotic cell expresses only a small portion of the entire genetic information at any time.

DNA SUPERHELICITY AND REGULATION

The regulation of negative superhelicity is known to modulate the initiation of numerous biologically important processes, including transcription, replication, recombination and repair.[28] The actively transcribing genes in eukaryotes are known to have an altered structure which renders them preferentially sensitive to nuclease digestion.[20] Experiments have shown that continued topoisomerase II enzyme activity is required to maintain the DNA in this transcriptionally active state. Inhibitors of this enzyme reverse both the activation of the genes and the nuclease hypersensitivity. The experimental evidence indicates that this altered structure at transcriptionally active sites is maintained by the DNA superhelicity resulting from continuous topoisomerase activity, and that relief of (or failure to maintain) this superhelical tension causes the DNA to revert to a less sensitive and less active state. For example, high rates of gene expression are observed from superhelical DNA templates that have been injected into frog oocytes.[17] Transcription effectively ceases when restriction enzymes are subsequently injected which linearize the DNA.

The specific activities involved in eukaryotic DNA regulation, and the role of transition-susceptible sites in them, are best studied using animal viruses of known sequence. The mammalian SV40 viral DNA is commonly used for this purpose. Because this virus co-opts the cellular machinery to effect its transcription and replication, the regulation of these events in SV40 presumably typifies eukaryotic regulatory controls in general. This is a much-studied virus whose base sequence is entirely known. Many of its regulatory sites have been well characterized.

SV40 has a closed circular genomic DNA. Inside the cell this molecule festoons itself with 24 nucleosomes, forming a structure called the SV40 minichromosome.[33] The locations of these nucleosomes is roughly uniform, although one region of the DNA remains free of them. Approximately 2% to 5% of the SV40 minichromosomes are transcriptionally active within a cell.[22] Significantly, the same proportion are superhelically stressed.

The sequence of the SV40 genome consists of two regions, a number of genes spanning about 90% of the molecule, and a single regulatory region of complex structure comprising the balance. This regulatory region includes the origin of replication, binding sites and promoters, and enhancer sequences. The 10% of the DNA which comprises this regulatory site is the nucleosome-free region of the minichromosome.

During the SV40 life cycle two sets of genes are activated. The early genes and the late genes are located on different strands of the molecule, hence transcribe in different directions. There are two early genes—the large T and small t antigens. The large T antigen gene consists of two exons. The large T antigen protein binds to sites around the three 21-bp tandem repeats, and thereby represses further transcription from these early genes. Moreover, its binding to site II at base pair 13 (on the edge of the inverted repeat), together with a topoisomerase activity, induce extreme negative superhelicity which activates replication.[33] The transition from early to late gene expression occurs after replication has been initiated.

The SV40 regulatory region contains two tandem copies of a 72-bp enhancer. This enhancer influences the initiation of both transcription and replication. Only one copy of the enhancer is sufficient to perform these functions, of which only the origin-proximal 30 bps are essential.[33] This includes an 8-bp Z-susceptible site, which is as long as any run of alternating purine and pyrimidine bases in the molecule. The 5099 base pairs that exclude the enhancers is conspicuously depleted of long Z-susceptible sequences, there being none longer than that found in the enhancer. Binding of chiral complexes indicates that this enhancer region exists in an altered, left-handed conformation which is probably Z-DNA.[23]

The functioning of the SV40 enhancer has been investigated by placing it in the pSV2 plasmid, together with its promoters.[34] Four forms of this plasmid were prepared—supercoiled, relaxed, nicked and linear. The plasmid was transfected into CV-1 or L-cells, and the level of gene expression was assayed. The level of transcription from the supercoiled plasmids was roughly two orders of magnitude higher than from the linear form. If the enhancer was deleted but the promoter retained, the effect of topology on gene expression was much less pronounced, making only a 3-fold difference in activity level. In fact the linear DNA with enhancer was 2- to 4-fold more effective than the supercoiled DNA without it.

The effect of the enhancer in supercoiled DNA does not depend on which side of the gene it is found on, or on which orientation it has relative to the gene. There is evidence that supercoiling does not influence the binding of enhancer-binding factors to enhancer sequences. Weintraub et al. conclude that enhancer-binding factors are present in excess and promoter-binding factors are limiting.[34] The enhancer activates the binding of factors involved in transcription to portions of the promoter. Also, enhancers appear to increase the efficiency with which these transcription factors are used. The rate-limiting step appears not to be the binding of these transcription factors, but the efficiency with which they are used. The imposition of negative superhelicity affects this efficiency of usage.

One model that explains all the experimental information regarding the role of these Z-susceptible enhancer sites posits that the molecule becomes activated by the imposition of superhelical stress. In this state the susceptible enhancer sites are transformed to Z-form. An enhancer-binding factor recognizes this alternative structure and in so doing induces a local environment in which this region is transformed back to B-DNA. This reversion couples to a destabilization of the promoter secondary structure, perhaps by strand separation, that is sufficient to increase the efficiency of initiation by some promoter-specific factor. For this mechanism to work, long Z-susceptible runs must not occur outside of the enhancer because they would compete with the enhancer sites for transition to the alternative conformation. This mechanism also explains the necessity for the non-enhancer portions of the SV40 genome to have the observed, statistically very significant, lack of long Z-susceptible sites.

ACKNOWLEDGMENT

This research was supported by grants DMB 86-13371 and DMB 88-96284 from the National Science Foundation.

REFERENCES

1. Bauer, W. R. "Structure and Reactions of Closed Duplex DNA." *Ann. Rev. Biophys. Bioeng.* **7** (1978): 287–313.
2. Benham, C. J. "Torsional Stress and Local Denaturation in Supercoiled DNA." *Proc. Nat'l. Acad. Sci. USA* **76** (1979):3870–3874.
3. Benham, C. J. "Theoretical Analysis of Transitions between B- and Z-Conformations in Torsionally Stressed DNA." *Nature* **286** (1980a):637–638.
4. Benham, C. J. "The Equilibrium Statistical Mechanics of the Helix-Coil Transition in Torsionally Stressed DNA." *J. Chem. Phys.* **72** (1980b):3633–3639.
5. Benham, C. J. "Theoretical Analysis of Competitive Conformational Transitions in Torsionally Stressed DNA." *J. Mol. Biol.* **150** (1981):43–68.
6. Benham, C. J. "Stable Cruciform Formation at Inverted Repeat Sequences in Supercoiled DNA." *Biopolymers* **21** (1982):679–696.
7. Benham, C. J. "Statistical Mechanical Analysis of Competing Conformational Transitions in Supercoiled DNA." *Cold Spring Harbor Symp. Quant. Biol.* **47** (1983):219–227.
8. Benham, C. J. "Theoretical Analysis of Conformational Equilibria in Superhelical DNA." *Ann. Rev. Biophys. Biophys. Chem.* **14** (1985):23–45.
9. Benham, C.J. "Superhelical DNA." *Comments Molec. Cell. Biophys.* **4** (1986):35–54.
10. Benham, C. J. "The Influence of Tertiary Structural Restraints on Conformational Transitions in Superhelical DNA." *Nuc. Acids Res.* **15** (1987):9985–9995.
11. Bianchi, M., M. Beltrame, and G. Paonessa. "Specific Recognition of Cruciform DNA by Nuclear Protein HMG1." *Science* **243** (1989):1056–1059.
12. Broydes, S., and D. Pettijohn. "Interactions of the *Escherichia coli* HU Protein with DNA." *J. Mol. Biol.* **187** (1986):47–60.
13. Eisenberg, D., and D. Crothers. *Physical Chemistry.* Menlo Park, CA: Benjamin-Cummings, 1979, 678–679.
14. Fuller, F. B. "Decomposition of the Linking Number of a Closed Ribbon: A Problem from Molecular Biology." *Proc. Nat'l. Acad. Sci. USA* **75** (1978):3557–3561.
15. Gierer, A. "Model for DNA and Protein Interactions and the Function of the Operator." *Nature* **212** (1966):1480–1481.
16. Haniford, D., and D. Pulleyblank "Transition of a Cloned $d(AT)_n$-$d(AT)_n$ Tract to a Cruciform *in Vivo*." *Nuc. Acids Res.* **13** (1985):4343–4363.
17. Harland, R., H. Weintraub, and S. McKnight. "Transcription of DNA Injected into *Xenopus* Oocytes is Influenced by Template Topology." *Nature* **302** (1983):38–43.
18. Jaworski, A., W. Hsieh, J. Blaho, J. Larson, and R.D. Wells. "Left-Handed DNA in Vivo." *Science* **238** (1987):773–777.

19. Johnston, B. H. "The S1-Sensitive Form of d(C-T)$_n$·d(A-G)$_n$." *Science* **241** (1988):1800–1804.

20. Larson, A., and H. Weintraub. "An Altered DNA Conformation Detected by S1 Nuclease Occurs at Specific Regions in Active Chick Globin Chromatin." *Cell* **29** (1982):609–622.

21. Lilley, D. M. "The Inverted Repeat as a Recognizable Structural Feature in Supercoiled DNA Molecules." *Proc. Nat'l. Acad. Sci. USA* **77** (1980):6468–6472.

22. Luchnik, A., V. Bakayev, I. Zbarsky, and G. Georgiev. "Elastic Torsional Strain in DNA within a Fraction of SV40 Minichromosomes: Relation to Transcriptionally Active Chromatin." *EMBO J.* **1** (1982):1353–1358.

23. Nordheim, A., and A. Rich. "Negatively Supercoiled Simian Virus 40 DNA Contains Z-DNA Segments within Transcriptional Enhancer Sequences," *Nature* **303** (1983):674–679.

24. Panayotatos, N., and R. D. Wells. "Cruciform Structures in Supercoiled DNA." *Nature* **289** (1981):466–470.

25. Panayotatos, N., and A. Fontaine. "A Native Cruciform DNA Structure Probed in Bacteria by Recombinant T7 Endonuclease." *J. Biol. Chem.* **262** (1987):11364–11368.

26. Peck, L., A. Nordheim, A. Rich, and J. Wang. "Flipping of Cloned d(pCpG)$_n$·d(pCpG)$_n$ DNA Sequences from Right- to Left-Handed Helical Structure by Salt, Co(III), or Negative Supercoiling." *Proc. Nat'l. Acad. Sci. USA* **79** (1982):4560–4564.

27. Rich, A., A. Nordheim, and A. Wang. "The Chemistry and Biology of Left-Handed Z-DNA." *Ann. Rev. Biochem.* **53** (1984):791–846.

28. Rowe, T., and L. Liu. "The Role of Superhelical Tension in DNA Structure and Function." *Comments Molec. Cell. Biophys.* **5** (1984):267–283.

29. Sinden, R., and D. Pettijohn. "Chromosomes in Living *Escherichia coli* Cells are Segregated into Domains of Supercoiling." *Proc. Nat'l. Acad. Sci. USA* **78** (1981):224–228.

30. Sinden, R. "Supercoiled DNA: Biological Significance." *J. Chem. Educ.* **64** (1987):294–301.

31. Singleton, C., J. Klysik, S. Stirdivant, and R. D. Wells. "Left-Handed Z-DNA is Induced by Supercoiling in Physiological Ionic Conditions." *Nature* **299** (1982):312–316.

32. Vinograd, J., J. Lebowitz, and R. Watson., "Early and Late Helix-Coil Transitions in Closed Circular DNA." *J. Mol. Biol.* **33** (1967):173–197.

33. Watson, J., N. Hopkins, J. Roberts, J. Steitz, and A. Weiner. *Molecular Biology of the Gene.* Menlo Park, CA: Benjamin-Cummings, 1987.

34. Weintraub, H., P. F. Cheng, and K. Conrad. "Expression of Transfected DNA Depends on DNA Topology." *Cell* **46** (1986):115–122.

35. White, J. H. "Self-Linking and the Gauss Integral in Higher Dimensions." *Am. J. Math.* **91** (1969):693–728.

Andrzej K. Konopka and John Owens
National Institutes of Health, National Cancer Institute, Laboratory of Mathematical Biology,
Bldg. 469, Rm. 151, Frederick, Maryland 21701

Non-Continuous Patterns and Compositional Complexity of Nucleic Acid Sequences

All properties of an organism are somehow encoded in its genome, a molecule of nucleic acid that is physically a linear copolymer of four monomers. Therefore, it can be represented as a sequence of four symbols and then simply treated as a text carrying many (generally unknown) messages written in many (also unknown) languages.

Experimental molecular biology provides evidence that the genome is non-uniform in the sense that its different regions are involved in different biological processes. Bacterial genomes primarily consist of genes coding for proteins and transfer and ribosomal RNAs (if we neglect rare instances of repetitious DNA: ribosomal genes, transposons and insertion elements). In contrast, genomes of multicellular eukaryotes virtually consist of repetitious DNA: introns, pseudogenes, highly repetitive regions (so-called simple sequence DNA), spacers between transcription units, transposable elements, and so on. Only a small percent of all nucleotides in these genomes form exons that encode proteins.[1,5,2,7,8]

This paper is devoted to scientific aspects of sequence research. Therefore strictly computational issues will not be addressed here. The subject of our concern is an answer to the question of whether the above-mentioned putative domains in nucleic acids can be characterized by differences in the 'non-randomness' of their sequences. In this connection a measure of local compositional complexity will be defined and then applied to an exemplary analysis of exons, introns and bacterial genes. To our knowledge compositional complexity at the level of short oligonucleotides has not been measured before in a quantitative manner and therefore all results concerning complexity can be considered as essentially new.

The method of non-contiguous pattern analysis[3,4] will also be considered. The basic idea of this methodology is to study pairs of short oligonucleotides separated by a gap instead of studying contiguous runs of nucleotides. This method is particularly useful for detecting the existence of code-like patterns in a collection of functionally equivalent sequences and the average length of the code word. Here we demonstrate once again that the 2-base periodicities in introns are caused by clustering of dinucleotides. We suggest that this might be a reason why introns display lower compositional complexity than exons.

COLLECTIONS OF SEQUENCES ANALYZED

Many entries present in the data banks are just variants of the same sequence. Also many sequences are nearly identical because they represent a similar function in organisms that are closely related (as all papovaviruses, adenoviruses and constant regions of immunoglobulin genes). For this reason, one cannot rely on automatic extraction from a data bank. An extracted file of sequences has to be 'cleaned' in order to represent a correct sample suitable for statistical analysis. All sequences used in this study were extracted from the GenBank (release, 58, Dec. 1988). Nearly identical sequences were then deleted from the collections by using a program CLEANER that makes use of the fact that a given long oligonucleotide (we used 15-mers) is expected in at most one sequence from a collection even as large as the entire GenBank (the program is available from the authors as well as from C. Burks at LANL). 'Clean' collections of sequences used in this study were:

- Multicellular eukaryotic introns (320 sequences, 121,858 nucleotides)
- Multicellular eukaryotic exons (522 sequences, 660,625 nucleotides)
- Bacterial protein coding genes (290 sequences, 300,597 nucleotides)

LOCAL COMPOSITIONAL COMPLEXITY

The measure of complexity used in this paper originates from the statistical theory of signal transmission.[6] According to it, compositional complexity of an oligonucleotide of length L is equal to the value of entropy function given by the formula:

$$H = -(1/L)^*[N1^* \log(N1/L) + N2^* \log(N2/L) \\ + N3^* \log(N3/L) + N4^* \log(N4/L)]$$

where L stands for the length of the oligonucleotide under consideration and $N1$, $N2$, $N3$ and $N4$ stand for the number of A, C, G and T characters respectively. The log function is taken to the base 4, thereby allowing the H values to be conveniently located between 0 and 1.

This definition implies that a homopolymer track will be the simplest $(H = 0)$ and an oligonucleotide that consists of 25% of each nucleotide will be the most complex $(H = 1)$.

In order to determine if eukaryotic sequences tend to employ simple short oligonucleotides more frequently than complex ones, regression analysis of the probability of octanucleotides as a function of their complexity was performed. For every octanucleotide in every collection of sequences, the probability of occurrence P and complexity H were determined. The expected probabilities $Pexp$ were calculated for hypothetical sequences that contain 25% of each nucleotide and the scores of occurrence $F = (P - Pexp)/Pexp$ were determined. All data records (the record consisted of complexity, actual probability, expected probability, and F-score) were then sorted in the decreasing order of F and average values of Hav and Fav for every group of 1000 records were calculated. Finally routine regression analysis of occurrence F as a function of complexity H was performed on such grouped data.

It can be seen from Figures 1a and 1b that in eukaryotic exons and introns, complexity H of octanucleotides is significantly correlated with their F-score of occurrence. The same effect takes place for tetra- through heptanucleotides (data not shown). The fact that the F-score of occurrence tends to decrease with the increase of complexity suggests that eukaryotic DNA 'employs' simple (and 'avoids' complex) oligonucleotides in a systematic manner. In bacterial genes (see Figure 1c), the F-score of occurrence is not correlated with complexity, which suggests that simple and complex oligonucleotides are employed with comparable frequencies. This confirms, in a quantitative manner, the suggestion[8] that eukaryotic DNA is 'cryptically simpler' than bacterial DNA at the level of short oligonucleotides.

Another observation concerns the average complexity of short oligonucleotides. For octanucleotides in exons and bacterial genes, it is equal to 0.823 and 0.839, respectively. The standard deviations of these means are 0.012 and 0.031, respectively. Therefore, mean complexity of exons and bacterial genes is the same within the error limits. The fact that octanucleotides in introns are significantly simpler than in exons and bacterial genes (mean complexity: 0.785 and standard deviation: 0.015) suggests that compositional complexity could be used for mapping introns in indiscriminately sequenced DNA fragments.

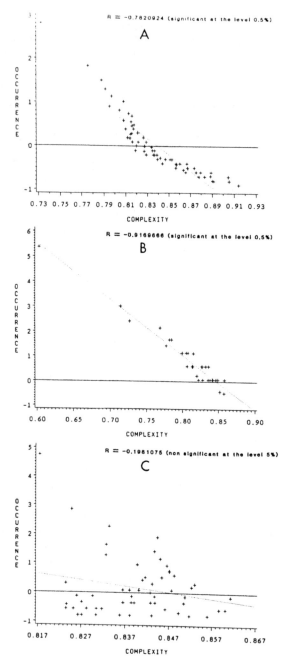

FIGURE 1   Results of regression analysis of the F-score of occurrence of octanucleotides (see the text) as a function of their compositional complexity for (A) exons, (B) introns and (C) bacterial protein-coding genes.

NON-CONTINUOUS, TWO-BASE PERIODIC PATTERNS IN INTRONS

The question that could be asked at this point is why exons display generally higher complexity than introns. One partial answer would be the occurrence of different code-like patterns in both kinds of sequences. Figure 2 presents examples of a distance chart[3] for mirror-symmetric trinucleotides in introns and exons. It can be seen from Figure 2A that the preferred gap length between two occurrences of mirror-symmetric trinucleotides in exons is $0, 3, 6, 9, \ldots$ and generally $3 * N (N = 0, 1, 2, 3, 4, \ldots)$; that is, these trinucleotides occur in exons in a 3-base periodic manner. Analysis of shortest distances (Figure 2B) shows however that this 3-base periodicity is not caused only by frequent occurrence of tandem repeats. Figure 2C shows that, in introns, the same trinucleotides tend to be repeated at the distances $1, 3, 5, 7, \ldots$ and generally $1 + 2 * N$; that is, they form 2-base periodic patterns. Moreover, Figure 2D shows that this periodicity is caused by a predominant clustering of mirror-symmetric trinucleotides at a distance of 1. A more exhaustive study of 2-base periodicities in introns and a detailed description of DISTAN methodology have been published elsewhere.[3,4]

In order to study the relationship between sequence periodicity and complexity, we explored a notion of *repetitive complexity* (that is, overlapping with but different from the previously discussed compositional complexity). Probabilities of repeats (of any length) for mono- through tetranucleotides at the distances 0 to 10 were calculated by taking a proportion of the number of all bases involved in repeats to the total number of bases in the individual sequence. Then the mean value P of these probabilities (over all sequences) was determined and the repetitive complexity $RC = 1 - P$ was calculated for every kind of repeat studied (mono- through tetranucleotides at the distances 0 to 10). Expected values of $Pexp$ and the corresponding standard deviations S were calculated for random sequences that preserve the composition of oligonucleotides under consideration. The expected repetitive complexity $RCexp = 1 - Pexp$ was then determined and the z-score, $z = (RC - RCexp)/S$, calculated.

Table 1 shows z-scores of Δrepetitive complexity for all mono- through tetranucleotides at the distance from 0 to 10 nucleotides in introns, exons and bacterial genes. As can be seen from Table 1, mononucleotide z-scores are all positive numbers. This suggests that at the mononucleotide level all sequences studied are more complex than corresponding random sequences. It can be seen as well that in exons and bacterial genes, trinucleotides favor distances $3 * N$ (more negative z-scores for distances $0, 3, 6$ and 9 than for $1, 2, 4, 5, 7, 8,$ and 10), tetranucleotides $1 + 3 * N$, dinucleotides $1 + 3 * N$ and mononucleotides $2 + 3 * N$. Therefore we can conclude that all short oligonucleotides tend to occur in exons in a 3-base periodic manner. The same kind of positional correlation confirms widespread 2-base periodicity of dinucleotides in introns (see Table 1) as underlined negative z-scores show that complexity is lower at the distances $2 * N$ than at $2 * N - 1$ and $2 * N + 1$.

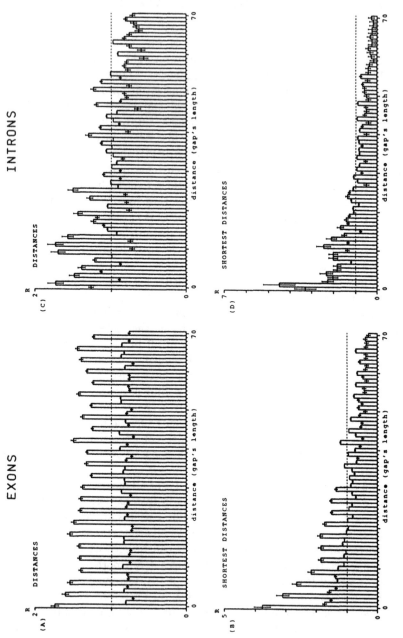

FIGURE 2 Distance profiles of mirror-symmetric trinucleotides provide evidence for the 3-base periodicity in exons (A and B) and for the 2-base periodicity in introns (C and D). R is the ratio of frequency of a given gap's length to the corresponding expected frequency. The error bars correspond to standard deviations from mean values of R over all sequences analyzed. The dashed line indicates that for a "random" sequence one should expect R=1.

TABLE 1 z-Scores of Complexity Due to Distant Repeats for All Mono- through Tetranucleotides in Introns, Exons and Bacterial Genes[1]

| | Distance | | | | | | | | | | |
	0	1	2	3	4	5	6	7	8	9	10
Eukaryotic Introns											
mo	2.4	2.3	3.1	2.6	2.4	2.7	3.1	2.9	3.3	2.9	3.2
du	−1.6	−0.9	−1.6	−1.2	−1.9	−0.9	−1.2	−0.9	−1.2	−0.8	−1.0
tr	−2.3	−2.7	−3.1	−2.9	−2.2	−2.6	−2.3	−2.5	−2.3	−1.8	−1.8
te	−1.9	−3.1	−2.3	−2.1	−2.2	−1.6	−2.4	−2.2	−1.8	−1.2	−1.3
Eukaryotic Exons											
mo	2.4	3.0	2.0	3.4	3.4	2.5	3.4	3.5	3.0	3.9	3.9
du	0.1	−1.8	0.5	0.0	−1.3	0.1	−0.1	−1.9	0.4	0.8	−0.6
tr	−2.4	0.0	0.0	−1.7	−0.2	−0.3	−3.9	−0.2	0.1	−1.2	0.0
te	0.8	0.8	−0.7	0.4	0.0	−2.6	0.8	1.1	−0.4	0.5	0.8
Bacterial Genes											
mo	2.7	3.9	2.3	3.7	3.8	2.5	3.6	3.3	2.1	3.0	3.0
du	−0.2	−1.4	−0.4	−0.3	−1.3	−0.4	−0.4	−1.3	−0.3	−0.3	−1.2
tr	−1.9	−0.6	−0.5	−1.7	−0.6	−0.6	−1.7	−0.5	−0.5	−1.7	−0.6
te	−0.1	0.0	−1.0	−0.1	−0.1	−1.0	−0.1	−0.1	−1.0	−0.1	−0.1

[1] Repetitive complexity at the level of mononucleotides is higher in the sequences studied than in corresponding random sequences (positive values of z-score at all distances). At the levels of di-, tri- and tetranucleotides, repetitive complexity is lower than expected (negative values of z-score) in most cases. Underlined negative numbers indicate the 2-base periodicities of dinucleotides in introns and the 3-base periodicities of trinucleotides in translated regions of genes (exons and bacterial genes).

A conclusion that can be drawn from the above results is that introns and exons might carry different messages encoded in a different way. In the case of exons we know the encoding schema and the code itself. In the case of introns we do not. Nevertheless, it seems likely that 2-base periodicities in introns are the cause of their low complexity. If sequences that display 4-base periodicity (or clustering of all tetranucleotides) would ever be discovered, they would probably be more complex than exons and bacterial genes. The functional significance, if any, of the 2-base periodicities observed in introns remains to be explained.

ACKNOWLEDGMENTS

We wish to thank Jacob V. Maizel for his constant enthusiasm toward studying non-contiguous patterns as well as for multiple discussions of the topic.

REFERENCES

1. Britten, R. J., and D. E. Kohne. "Repeated Sequences in DNA." *Science* **161** (1968):529–540.
2. Gall, J. G. "Chromosome Structure and the C-Value Paradox." *J. Cell Biol.* **91** (1981):3s–14s.
3. Konopka, A. K., and G. W. Smythers. "DISTAN—A Program which Detects Significant Distances between Short Oligonucleotides." *CABIOS* **3** (1987):193–201.
4. Konopka, A. K., G. W. Smythers, J. Owens, and J. V. Maizel. "Distance Analysis Helps to Establish Characteristic Motifs in Intron Sequences." *Gene Anal. Techn.* **4** (1987):63–74.
5. Long, E. O., and I. B. David. "Repeated Genes in Eukaryotes." *Annu. Rev. Biochem.* **49** (1980):727–764.
6. Shannon, C. E. "A Mathematical Theory of Communication." *Bell System Tech. J.* **27** (1948):379–423.
7. Singer, M. F. "SINES and LINES: Highly Repeated Short and Long Interspersed Sequences in Mammalian Genomes." *Cell* **28** (1982):433–434.
8. Tautz, D., M. Trick, and G. A. Dover. "Cryptic Simplicity in DNA is a Major Source of Genetic Variation." *Nature* **322** (1986):652–656.

A. Lapedes,† C. Barnes*, C. Burks,† R. Farber,† and K. Sirotkin†

†Theoretical Division, Los Alamos National Laboratory, Los Alamos, NM 87545 and
*Applied Theoretical Physics Division, Los Alamos National Laboratory, Los Alamos, NM 87545

Application of Neural Networks and Other Machine Learning Algorithms to DNA Sequence Analysis

In this article we report initial, quantitative results on application of simple neural networks and simple machine learning methods, to two problems in DNA sequence analysis. The two problems we consider are:

(1) Determination of whether procaryotic and eucaryotic DNA sequences segments are translated to protein. An accuracy of 99.4% is reported for procaryotic DNA (*E. coli*) and 98.4% for eucaryotic DNA (*H. Sapiens* genes known to be expressed in liver).

(2) Determination of whether eucaryotic DNA sequence segments containing the dinucleotides "AG" or "GT" are transcribed to RNA splice junctions. An accuracy of 91.2% was achieved on intron/exon splice junctions (acceptor sites) and 94.5% on exon/intron splice junctions (donor sites).

The solution of these two problems, by use of information processing algorithms operating on unannotated base sequences and without recourse to biological laboratory work, is relevant to the Human Genome Project. A variety of neural network, machine learning, and information theoretic algorithms are used. (For the purposes of this article, we view neural networks solely as an information processing procedure and do not consider the possible relation of these formal models to biological networks of neurons.) The accuracies obtained exceed those of previous investigations for which quantitative results are available in the literature. They result from an ongoing program of research that applies machine learning algorithms to the problem of determining biological function of DNA sequences. Some predictions of possible new genes using these methods are listed—although a complete survey of the *H. sapiens* and *E. coli* sections of GenBank using these methods will be given elsewhere.

INTRODUCTION

Annotated DNA sequence data presently available in databases such as GenBank[6] provide an important source of data for evaluating machine learning algorithms, statistical techniques, and information theoretic methods that may be useful in the Human Genome Project. When larger amounts of data are available from the Human Genome Project,[17] it will be essential to have automated, accurate, and verified symbolic information-processing techniques that help sift the raw sequence data for biochemically important features and to reduce the amount of laboratory work required to answer questions of biological interest.[9] The value of such techniques depends on their accuracy, which may be determined only by careful quantitative analysis.

In this article we use relatively new ideas from the field of neural network research (in combination with other machine learning algorithms, statistical analysis, and information theory) to address two problems:

1. Whether procaryotic and eucaryotic DNA sequence segments are translated to protein, and
2. Whether eucaryotic DNA sequence segments containing the dinucleotides "AG" or "GT" are transcribed to RNA splice junctions.

There is a wide literature[18] on applications of statistics to genetic analysis, and some of the neural net techniques we use are related to certain types of statistical approaches. In particular, we note that simple statistical weighting schemes commonly used in DNA analysis[34,35] are a subset of general neural net approaches, and therefore neural nets may be more general and more useful. Although we are able to decipher the way the neural net arrived at the results for the first problem we consider here, this is not always possible; and application of "automatic rule-based" systems provide an alternative method by which the explicit rules for arriving at an answer are inferred. Finally, we describe information theoretic techniques to produce graphs illustrating where in the DNA sequence the information relevant to solving Problem 2 resides.

The approach advocated here, i.e., quantitative analysis of various new (and older) information-processing methodologies, can be used to judge which algorithms might best play a practical role in the analysis of genetic data. We describe algorithms that we found to have significant success and present quantitative verification of accuracy. Due to the sparsity of rigorous quantitative statements concerning other approaches in the literature, only rarely were we able to compare the new algorithms to other methods. We found significantly improved accuracies for the two problems considered here. Extensive surveys of GenBank will be presented elsewhere.

GENERAL METHODOLOGY AND SPECIFIC DNA SEQUENCES USED

PROBLEM 1

The first problem deals with isolated sequence segments of DNA (ranging in length from 15 to 270 bases). We address the question of whether the segment contains sufficient local information to determine if it is translated to protein. We consider DNA from *E. coli* and also from *H. sapiens*. The problem is virtually identical to one considered by J. Fickett[12] in a careful investigation using different methods. Accuracies reported here are significantly higher—although we caution that due to the variability of genetic data, comparisons of accuracies are best performed on identical data sets. We did not have Fickett's data available.

Data from GenBank release number 57.0 were used in the following manner for *H. sapiens:* A "training set" of fragments was constructed from two subsets of data. The "true" subset consisted of in-reading-frame segments of DNA of fixed length taken from known coding regions. The "false" subset consisted of segments of identical length annotated in GenBank as not coding for protein—selected at random. (For *H. sapiens* the "false" set was selected solely from segments contained in annotated introns.) A DNA segment in the false set can contain an "in-frame" stop codon (relative to a "frame" defined by the start of the segment), and hence this could provide a trivial distinction between segments from the "true" set (with no in-frame stop codons) and segments from the "false" set. Thus the "false" segments had any stop codons in them removed to avoid the obvious bias of classifying any segment with an "in-frame" stop codon as a member of the false set. We emphasize that we therefore deal only with classifying Open Reading Frames (ORF's). To eliminate stop codons in the "false" set, we investigated two alternative methods. The first method, which was used to generate the data presented in this manuscript, eliminated an entire segment of DNA if a stop codon appeared in a segment of the false set. That is, if in the process of selecting segments of a specified length from introns, a stop codon appeared "in-frame" in that segment (relative to the start of the segment), then that segment was not used in forming the "false" set. The second method merely excised the stop codon and replaced it with the next succeeding codon. Interestingly, accuracy seemed to differ significantly depending on what method of dealing with stop codons was used (not shown). These true and false subsets comprise the "training" set.

For *H. sapiens* the "true" set came from regions annotated as coding ("pept") and being expressed in liver ("liver cDNA" in SOURCE field). Pseudo-genes and duplicate members from multigenes were removed. The "false" set was prepared similarly, except from introns ("IVS"), and tissue type was ignored.

The first approximately 50 non-duplicated *E. coli* GenBank entries from an older release of the database were processed as follows. Regions annotated as coding for protein were taken for the "true" set. Regions neither coding for protein nor coding for structural RNAs were taken for the "false" set. The "true" and "false" sets for the training procedure for *E. coli* were constructed by extracting the first 400 bases of every other sequence from the above sequences. These data were then segmented into lengths of 15-, 45-, 90-, 135-, 180-, and 270-base-long fragments in order to test the effects of segment length. The prediction set was the rest of *E. coli*. After training is complete one may "window" a new sequence into appropriate segments and attempt identification of protein-coding regions.

Data from the training set were used to fix any adjustable parameters in the various algorithms. A disjoint "prediction set" of true/false segments was formed in a similar manner to the training set from sequences not used in constructing the training set. After the training set was used to adjust parameters appearing in the algorithm, the algorithm was evaluated on the prediction set. Evaluation of the algorithm on the prediction set, which consists of data not used in construction of the algorithm, is indicative of the accuracy of the algorithm on new data produced by experiment. Our results for Problem 1 were reproduced to within 0.1% upon different random partitions between training and prediction sets, when the size of the partitions is the same between experiments.

[If new, experimentally produced data is not representative of the known selection effects in current GenBank data (e.g., over-representation of genes coding for abundant proteins), then accuracy on the prediction set may not be indicative of the accuracy on new data. This is an unavoidable consequence of current experimental techniques. In a more detailed analysis, one may wish to train networks on subclasses of GenBank data that may then generalize to special situations.]

We point out that the methodology of dividing the data into a "training set" and "prediction set" is critical for obtaining meaningful results. Prediction on the training set (and not on new data) can lead to gross overestimates of accuracy, particularly when data sets for training are limited. This is particularly evident for Problem 2 (see Tables 3, 4, 5, and 6 in the Results section). Numerous papers in the literature use training data in attempts to evaluate the accuracy of algorithms. The method of eliminating "in-frame" stop codons from the false set can also affect results, as well as the size of the training set. (Therefore, care must be taken in comparing the results of algorithms used here to results present in the literature.)

PROBLEM 2

The second problem deals with the determination of functional splice sites in eucaryotic DNA. The analysis used a similar methodology to the previous problem. A training set and a separate prediction set were constructed. The training set and prediction set each separately contained true and false examples. The training and prediction sets were constructed from Release 57 of GenBank. First the known (true) splice junctions were extracted along with a surrounding window of bases of specified length from both sides of the highly conserved dinucleotide. This was done separately for the boundary between an intron and an exon (acceptor sites, conserved dinucleotide = "AG") and for the boundary between an exon and intron (donor sites, conserved dinucleotide = "GT"). We did not consider the rare cases of deviation from these dinucleotides. This procedure was followed three separate times in order to test the effects of three lengths of the surrounding window of bases: a total of 11 bases, 21 bases, and 41 bases.

The "true" training sets and "true" prediction sets were constructed by first randomly shuffling the order of these true examples and then selecting all but 50 of the examples to be a true training set for each length category, while the remaining 50 were the true prediction set for that category. The reason for using a majority of data in the training set is that the total amount of presently available splice junction data is limited. Accuracy was increased by using more of the data during training to accurately determine parameters found in the algorithm. The resulting small size of the prediction set was then compensated for by repeating the random division into training sets and prediction sets a number of times and averaging the accuracies of the resulting networks. The results we report are averaged over five trials. This method is a variant of the classic "leave one out" procedure commonly used in applications of machine learning algorithms to small data sets.[11]

The false sets were constructed by similarly extracting 11-, 21-, and 41-base windows surrounding the dinucleotides, "AG" (false acceptor splice sites) and "GT" (false donor sites), occurring inside exons. For each length category, the number of false examples in the training set far exceeded the number of true examples. This overemphasis on the false category examples can be compensated for in the neural net training algorithm. In a separate investigation we studied the effect of choosing examples of the false set from windows surrounding AG or GT in both exons and introns (as opposed to selecting the false set from windows surrounding AG or GT that are only in exons). Here the number of false examples was chosen to equal the number of true examples. Little variation in results was found between choosing the false set from both introns and exons, or solely from exons.

Even when care is taken to divide the data into separate predict and train sets, one finds for Problem 2 that identical patterns can occur between the train and predict sets. This is due to the highly restricted nature of the base composition surrounding the splice sites. A small, but not insignificant increase in accuracy may therefore be obtained by mere table look-up. In the future, when one knows whether the duplicated patterns are representative of a true prototypical class, one may wish to capitalize on this increase in accuracy. In this paper we present results on non-overlapping train and predict sets in order to obtain insight into the ability of the algorithms we use to generalize to new situations. The results we present are therefore presumably a conservative, lower bound on predictive accuracy. Results on incorporating the increased table look-up accuracy (best done by "rule-based" systems) will be reported elsewhere.

We note that "clean" determination of false splice sites is complicated by the possibility of alternative splicing. In this preliminary study, to avoid complications due to alternative splicing, we removed genes that are prone to alternative splicing (immunoglobulins, oncogenes, etc.). Certainly other constructions of a training set are possible. The procedure used, however, does seem to be a reasonable "first cut" at the problem.

ALGORITHMS

A variety of methods were tested, all of which have their genesis in various approaches to machine learning. A great deal of knowledge concerning these algorithms is available in the literature, and their behavior on numerous problems involving the processing of symbolic informations is a focus of continuing research.[10,31,32]

NEURAL NETWORKS

Important work of Hopfield and others[10,15,31,32] has caused a resurgence of interest in neural net formalisms in the past five years. A classic neural net algorithm, the Perceptron Method,[11] is the forerunner of many neural net algorithms in use today. This method has previously been used by Nakata, Kaneisha, and DeLisi[26] in an investigation of the splice site recognition (our Problem 2) and by Stormo, Schneider, Gold, and Ehrenfeucht[36] in attempts to determine ribosome binding sites. Accuracies were not as high as could be desired. We first describe the general idea of the Perceptron Method and then discuss recent extensions[23,31] that have been shown to yield improved accuracies in many other interesting problems.

In the context of DNA analysis one starts by providing a suitable encoding of the symbols "A," "T," "G," and "C" into a string of bits, i.e., 0's and 1's. For example, each symbol can be represented by four bits, and these four bits/symbols are concatenated to provide a bit string representing a given length of DNA (see Figure 1). Other representations may be chosen, such as using 64 bits to represent each of the 64 possible codons.

A "neural network" can be constructed by assigning a "neuron" to each bit of the input string (with a state "on" or "off"), by assigning an additional neuron to be an output neuron, and by viewing the arrangement as a network where the input neurons influence the state of the output neuron. After training the network,

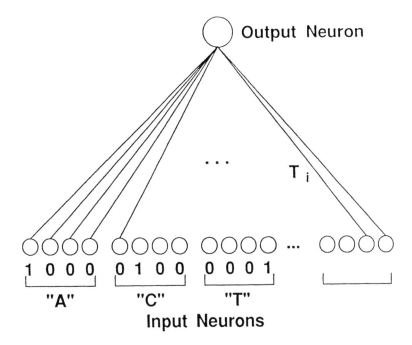

FIGURE 1   A first-order neural network, with one of the input representations
discussed in the text. Each symbol is represented by four bits. A string of symbols is
represented by concatenating the four bits for each symbol in the string. If a codon
representation instead of a base representation were used, each symbol (codon) would
be represented by sixty-four bits. The line of circles represents input neurons, which
take on values of 0 or 1. The output neuron state is determined by the input neuron's
states according to a set of connection weights and after training is complete should
attain a value near 1.0 if the input example comes from the "true" class; near 0.0
otherwise.

the output neuron should take on a value near one if the input string comes from
the true subset of the training set, and a value near zero otherwise (see Figure 1).
The state of the output neuron is given by

$$O = g\left(\sum_j T_j I_j + \theta\right), \tag{1}$$

where I_j represents the input values (0 or 1), O represents the state of the output
neuron, T_j are numerical "weights" to be determined by a learning algorithm, and
θ is an additional constant also determined by the learning algorithm. $g(x)$ is an
arbitrary smooth function, monotonically rising from the value 0.0 for negative
values of x, to the value 1.0 for positive values of x. We chose

$$g(x) = \frac{1}{2}\left(1 + \tanh(x)\right), \tag{2}$$

where tanh is "hyperbolic tangent," although other forms may be used without affecting results.

The weights are initialized to small random values. Training the network (i.e., finding the optimal T_j and θ) is accomplished as follows. Let p index the various examples in the training set, and let $t^{(p)}$ represent the target output value (0 or 1) for the pth input example. If the input example belongs to the true subset of the training data, the target output for that example should be a 1, and should be a 0 otherwise. Similarly, let $O^{(p)}$ represent the actual output of the net for the pth input example. The mean square error made by the network on the training set is given by

$$E = \sum_p \left(t^{(p)} - O^{(p)}\right)^2. \tag{3}$$

One clearly wishes to minimize the error, E. One way to do this is to make successive changes, ΔT_j and $\Delta\theta$, in the weights, T_j, and in θ by

$$\Delta T_j = -\frac{\partial E}{\partial T_j} \cdot \epsilon \tag{4a}$$

$$\Delta\theta = -\frac{\partial E}{\partial \theta} \cdot \epsilon, \tag{4b}$$

where ϵ is a small parameter. Simple algebra shows that the error will decrease to a minimum. We actually use an efficient conjugate gradient method[29] to minimize E, but Eq. (4) is often used in the literature, and it approaches the same result (albeit slowly).

After minimizing the error on the training set, the weights T_j and also θ will be known. One may then test the network's accuracy on the "predict set."

An alternative approach[4,16] is to view $O = g\left(\sum_j T_j I_j + \theta\right)$, which ranges between zero and one, as a probability of a discrete state output being a 1.0, given the input I. One can then compute the joint probability over the training set that all the examples from the true subset yield an output of 1.0 and that all examples from the false set yield an output of 0.0. This joint probability may be written as

$$P = \exp\left(\sum_p t^{(p)} \log\left(O^{(p)}\right) + \sum_p .(1 - t^{(p)}) \log\left(1 - O^{(p)}\right)\right). \tag{4c}$$

One may then define E to be the negative log of this probability,

$$E = -\log(P), \tag{4d}$$

and may minimize this quantity in a manner similar to Eq. (4). This is equivalent to maximizing the log of the probability P and therefore to maximizing the probability P. We refer to this method of training as the "maximum likelihood" method.

Rummelhart and McClelland[31] have recently extended the input/output layer formalism by including additional layers of neurons between input and output that may be viewed as a kind of intermediate processing layer. These neurons are conventionally called "hidden neurons" (i.e., neither accessible as outputs or as inputs) and considerably extend the power of the network in Figure 1. The outputs of the hidden neurons, given by Eq. (1), are now fed as inputs to the final output neuron. The equations become more complicated, but the form of Eq. (4) that is use 1 to find the weights remains the same. Details may be found in Rummelhart.[31] This algorithm is commonly called the back-propagation neural net method and is quite popular.

Another modification of the network has been suggested by Y. C. Lee et al.[23] (see also Rummelhart's sigma-pi units[31]). Eq. (1) is modified so that the output is no longer a function of the linearly weighted sum of inputs. Instead, a linearly weighted sum of products of inputs is used. One may use products of any desired order; however, one is often limited by the availability of data used to determine the additional weights or limited by computer resources. For the latter reason, we have not yet gone above third order. In this case Eq. (1) becomes

$$\mathbf{O} = g\left(\sum_{ijk} T_{ijk} I_i I_j I_k + \theta\right) \tag{5}$$

The analogue of Eq. (4) (with the appropriate extra indices) is used to find the T_{ijk}. This method is called the high-order correlation neural net method.

This method is closely related to classic statistical techniques of multi-variate analysis. Also, all of the techniques above are a form of discriminant analysis, and relations may therefore be found to classic statistical and discriminant techniques.[20,22] Note that if I_j represents bases, then Eq. (5) includes the effects of base correlations up to third order. Therefore, Eq. (5) can subsume the correlations among bases due to codons and to other base correlations as well. One advantage of the more powerful network formalisms is that they are capable of automatically forming appropriate "internal representations" of the data. For example, if the R, Y representation (i.e., the purine, pyramidine representation of nucleic acid bases) is more useful for solution of a particular problem, then a base representation may be used on the input layer, and an internal R, Y representation (or other appropriate representation) can be formed automatically by the network on the "hidden" layers. Similar "internal" representations may be found automatically by high-order neural networks. Decoding what internal representation was chosen by a network is in general a difficult problem, but it is one of obvious biological importance. We used Eq. (5) for the analysis of Problem 1 and were able to distinguish the effects of codon usage from other base correlations in a controlled fashion.

As we have seen, there are a variety of neural net formalisms, the simplest of which are reminiscent of the simple statistical weighting schemes previously used in DNA analysis. The power of the new neural net formalisms may also be understood as follows. It is possible to show that the simplest weighting scheme, Eq. (1), (with no "hidden" neurons), attempts to discriminate the true class from the false class by passing a separating hyperplane between the two classes.[11] This plane exists in a high-dimensional space, the dimensionality of which is related to the number of input neurons used to represent the data. The learning algorithm, Eq. (4), attempts to move the plane so that true patterns in the training set fall on one side of the plane and the false patterns on the other. A similar interpretation exists for attempts to adjust the weights in simple statistical schemes used in the biological literature (except that optimal procedures are not used to adjust the plane). However, it is easy to construct situations in which the two classes (true and false) do NOT fall on opposite sides of a plane, and one suspects that this situation is generic (see Figure 2). The extension of Eqs. (1) and (4) using hidden neurons, or Eq. (5) using high-order networks, allows curved separating surfaces to be used,[22,24] and one can see from Figure 2 that this is exactly what is required in the generic situation. This is one way in which the new neural net approaches achieve additional power over the older Perceptron approaches and related simple statistical weighting schemes.

However, it is important to understand that one can use less powerful networks (e.g., Perceptrons) with heavily pre-processed input patterns to acheive any result the newer algorithms produce. That is, there is a trade-off between putting processing power into the network versus putting it into a pre-processing stage that changes the representation of the input data. In terms of Figure 2, these two situations correspond to (1) using a powerful network to derive an appropriate curved surface separating the two classes versus (2) using a minimal network capable of producing only planar separating surfaces, but pre-processing the input data so that the separate class regions are deformed into convex shapes that can be separated by a plane. When one has some idea of the appropriate pre-processing (e.g., using an input representation for codons instead of single bases for Problem 1), then the latter procedure is useful. However, when one has little idea of the correct pre-processing (which is the more usual situation), then turning the job of classification over to a powerful net capable of automatically forming curved separating surfaces is useful. Clearly, the best approach is to encode as much relevant information as possible about the problem into the input representation and then to use a sufficiently powerful network to perform accurate classification. In fact, we concluded that a representation using codons (as opposed to bases) is useful for Problem 1 by first using a powerful network (a third-order net) on a base representation and then deducing that the network was responding most strongly to third-order base correlations over adjacent bases (i.e., codons). Thus in certain situations, one may deduce an optimal input representation by analyzing more powerful networks and then return to using the less powerful networks that operate on the optimal input representation. Appropriate input representations can greatly reduce training times.

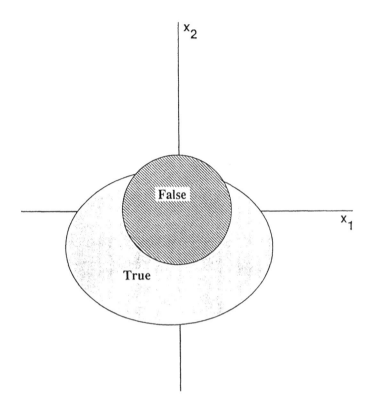

FIGURE 2 A two-dimensional example of "true/false" classes that can not be
separated by a line. In higher dimensions this would correspond to classes that could
not be separated by a hyperplane.

AUTOMATIC RULE-BASED METHODS, METRIC METHODS, AND INFORMA-TION THEORY

Automatic rule-based methods[14,27,30] are a class of methods that attempt to infer
an algorithm from a set of examples (as do neural nets) but operate in a much
different fashion from neural networks. Typically, at the end of a training procedure
for rule-based systems one obtains a collection of rules such as

$$\text{if(condition on symbols)} \rightarrow \text{(outcome)}.$$

Chaining these rules together provides a logical route to go from the symbolic
input data (e.g., a string of A, T, G, or C's for DNA) to a conclusion, e.g., "true
splice junction" or "false splice junction." One advantage of rule-based systems
is that the logic involved in arriving at a conclusion is transparent; whereas for
neural networks, the logic is encoded in the network architecture and the network

weights. The latter can sometimes make biological interpretation difficult. The set of rules obtained from a rule-based system are similar to "expert systems"[7,38] used in artificial intelligence; however, the rules are derived automatically without recourse to an outside "expert," thereby avoiding the "knowledge acquisition bottleneck" problem of expert systems. Rule-based systems have enjoyed considerable success in various problems[14,27,30] and in certain interesting cases have been shown to exceed neural nets in accuracy.[2]

The particular rule-based system we employed is due to Quinlan.[30] The idea is to grow a decision tree from the training set samples and use this tree to classify the prediction set. The decision tree is grown by finding those base positions that contribute the most information towards the classification of an example as "true" or "false" (e.g., "true" or "false" splice junctions) and then finding the most informative bases in those bases positions. "Information" is used in a technical sense and refers to the "mutual information" of the base position and the category "true" or "false."[30] The result of the algorithm is a binary tree that descends from a root node and becomes increasingly specific as one descends down the levels of the tree. Each node contains a template that displays which bases are the most important in each base position and also contains a classification category of "true" or "false" for examples matching each template. To use the tree for prediction, one takes the DNA sequence segment to be classified and checks if one of the templates on the bottom level of the tree matches it. If a match occurs, the sample is classified with the class category associated with the matching template. If no match occurs on one level, then one goes up a level in the tree (hence using less specific templates) and checks for a match, repeating the procedure until a match is found (which is guaranteed). Because the accuracy of the rule-based system in the cases we considered did not exceed the accuracy of the neural networks, we will not give full details of the rule-based system here. Explicit details of all algorithms used, as well as a complete survey of the *H. sapiens* and *E. coli* sections of GenBank using the various methods outlined here, will be available in Barnes et al.[3]

As explained above, one advantage of the rule-based system is that the logic leading to the classification of true or false is transparent. When the rule-based algorithm outlined above was used on the splice site recognition problem, it was possible to find a relatively small set of rules classifying the splice junctions with almost the same accuracy as the neural net. An example of some of these rules is given in Figure 3.

Futher work along these lines, combined with efforts to reduce rule systems to the minimal number of rules, could result in a relatively small set of biologically interpretable rules that can correctly allocate putative splice junctions to the true or false class with high accuracies.

Other methods exist for determining where information resides in a base sequence that contributes to the sequence being designated as one class or another (e.g., true or false splice sites). Considerable attention has been paid to consensus sequences associated with splice sites.[25] Consensus sequences describe the observation that true splice sites have highly conserved dinucleotides ("AG" and "GT" for

Decision Tree: Acceptor Splice Junctions

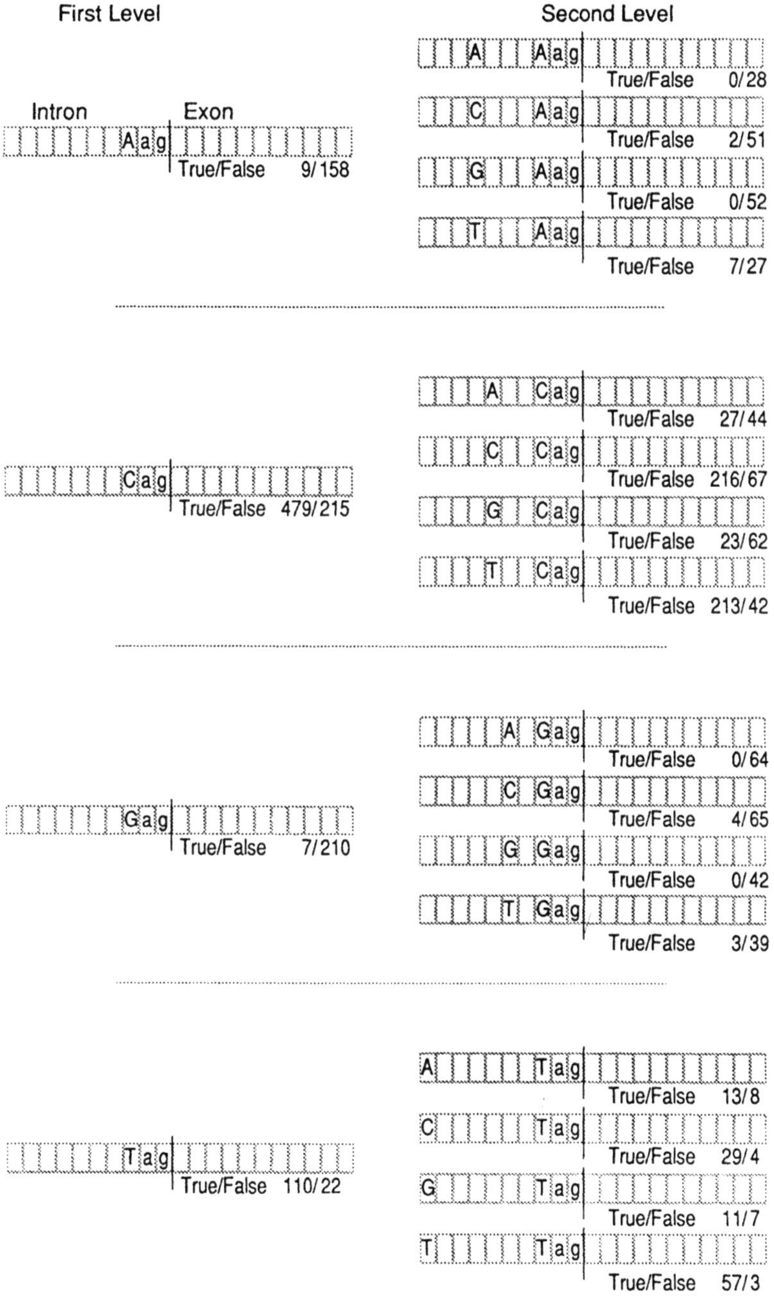

First Level

Second Level

FIGURE 3   Decision tree. (continued)

FIGURE 3   (continued) The figure represents the first two levels of a five-level decision tree constructed to solve the problem of recognizing acceptor splice junctions. The tree is constructed so that each deeper level reflects an increasing number of relevant base positions. The learning algorithm tests all base positions at a node to decide which is most relevant at that level. (For example, to go from the pattern shown at the top right-hand of the figure, ...A...A..., to the next lower level, the remaining 19 base positions would be tested, and the position with the highest information content would be selected.) The 21 boxes each represent base locations in the sequence upon which decisions may be based. The set of 21 taken together represents one window, or sequence segment. The vertical line denotes the position of the potential intron/exon boundary with the potential intron to the left. The AG consensus dinucleotide is shown in lower case. Since this dinucleotide is present in both "true" and "false" data sets, its base positions are not considered relevant by the learning algorithm. The upper case letters shown in the figure represent the bases at the most meaningful positions chosen by the algorithm. The numbers of examples from the true and false sets chosen by each decision rule are shown below the representation of each rule. At the first level the algorithm decided to partition on the eighth base. For example, when that base is an "A", the rule selected 9 true acceptor splice sites and 158 windows taken from non-splice-sites. Once the data are partitioned with an "A" in the eighth position, as shown at the top of the figure, the algorithm decides that the next most important position is the fourth position. When that is an "A," there are no actual examples from acceptor sites, but there are 28 examples taken from non-splice sites. In practice, the algorithms probe to deeper levels until either the true or false populations are either zero or acceptably small.

acceptor and donor splice sites, respectively) and in addition, in nearby positions, tend to have particular bases with significant frequencies. However, consensus sequences alone do not convey information that allows discrimination of true splice sites from false splice sites since they are constructed solely from base frequencies in the true set. If similar frequencies occur in the false set, then consensus sequences have little informative value. What is needed is a measure quantifying the differences in frequencies of bases surrounding true splice sites from those surrounding false splice sites. A suitable measure for comparing two such probability distributions is the Kullback divergence measure,[19] which in this context is a formula for the average number of bits of information in each base position contributing toward the decision true or false. The Kullback divergence measure is defined by $K(b)$ where

$$K_s(b) = \left(P_s^T(b) - P_s^F(b)\right) \log_2 \left(\frac{P_s^T(b)}{P_s^F(b)}\right) ; \tag{6a}$$

$$K(b) = \sum_{s=A,T,G,C} K_s(b) \tag{6b}$$

Here, b is an integer-labeling base position, and $P_s^T (b)$ represents the probability of symbol s (s = A, T, G, or C) occurring at base position b in the true set, and $P_s^F (b)$ is similarly defined for the false set.

The utility of such a measure is evident in Figure 4. In Figure 4 the true and false sets were those for the splice junction problem. Figure 4 clearly shows that the information due to single base positions contributing to the discrimination between true and false acceptor splice sites predominantly resides in roughly 15 bases on the intron side of an acceptor splice site. A similar plot (not shown) showed that the information contributing to the discrimination between true and false donor sites predominantly resides in four bases on the intron side of an donor site, but with additional information coming from approximately three bases on the exon side. When the window of bases surrounding the dinucleotides "AG" and "GT" were truncated to the widths mentioned above, we noticed that the multi-layer neural net method of prediction suffered only a small loss in accuracy, thereby

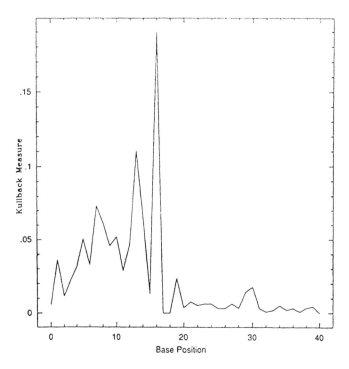

FIGURE 4   The Kullback information measure plotted as a function of base position for splice junction acceptor sites. This is a measure of the average number of bits of information in each base position that aid in discrimination of "true" from "false" sites. Base locations 17 and 18 contain the consensus dinucleotide AG, so the exon starts at location 19. Since AG occurs at positions 17 and 18 in all examples in both the "true" and "false" sets, the Kullback value at these positions is zero.

confirming that information in the regions described above is a key contributor to discrimination. Considerable discriminatory information residing in the intron side of splice junctions is in accord with the intron "lariat" mechanism proposed for SnRNA's during the intron excision.[13] The information residing on the exon side of donor splice junctions suggests that there may well be a biological correlate at work here as well.

Consensus sequences have also been used in attempts to represent the importance of base positions in splice sites. Typically such attempts display the most common base in a base position along with its frequency of occurrence in the "true" set. However, it is the *relative* probabilities in the "true" and "false" sets that determine the efficacy of the frequently occurring bases as a signal for a splice junction. This point is quite clear if one considers consensus sequences for the start of translation. Obviously "ATG" would appear with high probability, yet "ATG" also appears with high probability at other base positions not involved in the start of translation. Thus the value of the "ATG" as a signal for the start of translation is reduced due to the the "false" "ATG's" appearing elsewhere. The Kullback divergence quantifies the discriminatory power of a putative signal sequence.

The occurrence of significant information in a relatively few number of bases also suggests the use of other methods. A classic, and simple, machine learning method is the k nearest neighbor method.[11] One takes an element from the predict set; finds the k nearest neighbors in the training set using a suitable metric; and predicts the unknown pattern as being a true splice site if the majority of these nearest neighbors in the training set are true splice sites, and predicts false for the opposite case. The success of the method hinges on the choice of a "suitable" metric or distance measure. We investigated various metrics.

A simple Hamming metric for the distance between two base sequences may be defined as the number of base positions in which the two sequences have different bases. This measures the effect of mutations but can indicate a large distance if insertions or deletions of even a single base are involved. [The two sequences are prealigned so that the conserved dinucleotides (either "AG" or "GT") are in the same base position in both sequences.] A variant of this procedure is to weight each base position with the Kullback measure, so that if a discrepancy between the two sequences occurs in an unimportant base position [low K for base position (b)], then the contribution to the distance is minimal. We call this a "weighted" Hamming measure.

Effects of insertions and deletions (which can have radical effects on Hamming measures and also network methods) may be accounted for by using a k nearest neighbor method where the metric is that commonly used in sequence alignment.[33] We call this a "similarity metric" method. Note that if similarity to small nuclear RNA's (SnRNA's)[13] is important in discriminating true splice sites from false, then the k nearest neighbor similarity method should work well. We did not find this to be the case.

RESULTS

Neural net methods achieved the highest accuracies for Problems 1 and 2. High-order neural nets were used to determine the importance of codon, and also intercodon dinucleotides, by first utilizing a fifth-order net restricted to six adjacent bases. In later work we were then able to preprocess the input data into a "codon plus dinucleotide" representation and use a Perceptron. Results quoted here do not include the small, but significant, increase in accuracy obtained when the intercodon dinucleotide is included. Automatic rule-based methods achieved slightly lower accuracies for Problem 2 and were not applied to Problem 1. Hamming and weighted Hamming metric methods were consistently poor performers (70–80% accuracy on Problem 2), were not applied to Problem 1, and will not be reported in detail. These methods were included because they are relatively simple algorithms that have worked well in other investigations. Which method works best is generally problem dependent; thus, it is prudent to survey a number of methods when a problem is of particular interest. The similarity metric method (less sensitive to inserts and deletes) also performed roughly the same as the other metric methods and so will not be reported in detail either. Use of the Kullback divergence measure provides more useful information than consensus sequences, and the results are reported in Figure 4.

TABLES OF PERCENTAGE ACCURACY FOR PROBLEMS 1 AND 2 VERSUS DNA SEGMENT LENGTH

Percentage accuracy is defined as the average of two numbers. The first number is the ratio of the number of elements in the "true" prediction set, correctly labeled as true, to the total number of true elements in the prediction set. The second number is similarly defined for the "false" prediction set. The total percentage accuracy is the average of these two numbers, i.e., the average accuracy on both the true and false sets. The two percentage accuracies (percentage of true correctly called as true and the percentage of the false set correctly called as false) did not differ significantly when using neural net methods, and therefore the average is a useful expression of overall accuracy. The following tables result from using neural net methods. For *H. sapiens* the coding regions were limited to those known to be expressed in liver.

PROBLEM 1: DETERMINATION OF WHETHER OR NOT DNA SEQUENCE SEGMENTS ARE TRANSLATED TO PROTEIN.

TABLE 1 *E. coli*, Accuracy on Predict Set

DNA Segment Length	Percentage Accuracy
15 bases	76.1
45 bases	87.2
90 bases	93.8
135 bases	97.6
180 bases	99.5
270 bases	99.5

TABLE 2 *H. sapiens*, Accuracy on Predict Set

DNA Segment Length	Percentage Accuracy
45 bases	79.0
135 bases	91.9
270 bases	98.4

PROBLEM 2: DETERMINATION OF WHETHER OR NOT EUCARYOTIC DNA SEQUENCE SEGMENTS ARE TRANSCRIBED TO RNA SPLICE JUNCTIONS.

TABLE 3 *H. sapiens*, Acceptor Sites, Accuracy on Predict Set

DNA Segment Length	Percentage Accuracy
11 bases	85.3
21 bases	89.7
41 bases	91.2

TABLE 4 *H. sapiens*, Acceptor Sites, Accuracy on Train Set

DNA Segment Length	Percentage Accuracy
11 bases	85.0
21 bases	91.3
41 bases	96.0

TABLE 5 *H. sapiens*, Donor Sites, Accuracy on Predict Set

DNA Segment Length	Percentage Accuracy
11 bases	94.5
21 bases	94.5
41 bases	91.3

TABLE 6 *H. sapiens*, Donor Sites, Accuracy on Train Set

DNA Segment Length	Percentage Accuracy
11 bases	94.5
21 bases	95.7
41 bases	99.0

Accuracies obtained on Problem 1 compare favorably to Fickett in previous work on a related problem.[12] A minimum of 200 bases was required in this prior work, and an accuracy of 85% was achieved (previously quoted accuracies of 95% for Fickett's algorithm assume a 20% "no opinion" rate). Eighty-five-percent accuracy should be used to compare with our results because we do not use a "no opinion" category. We caution that the data sets used in the two algorithms were not identical and that this can affect a comparison of results.

Accuracies on Problem 2 exceed those for algorithms previously described in the literature such as the statistical method of Shapiro and Senapathy.[34] The Perceptron accuracies were significantly higher, which indicates the advantage of a procedure for optimally determining the weights. Nakata, Kaneisha, and DeLisi[26] have previously implemented a Perceptron algorithm on the splice junction problem. Again, our results are significantly higher, which we attribute to somewhat different methodologies, and also to the fact that Nakata et al. were limited by the small data sets available at the time of their investigation. A rough rule of thumb is that the number of training examples should be at least a few times the number of network weights in order for accurate predictions to be possible. Also note that accuracies on the training set are significantly higher than those achieved on the predict set and indicate that such an attempt to evaluate the accuracy of an algorithm yields significant overestimates.

A complete survey of the *E. coli* and *H. sapiens* sections of the GenBank database[6] using our methods will be presented elsewhere.[3] One may perform this survey by presenting the output of the networks in a graphical form that indicates coding regions and splice sites. An example of such a graphical description incorporating network predictions is in Figure 5. In the course of this investigation we did predict some potential protein-coding regions. One example[21] was the prediction of a previously unannotated protein-coding region in the GenBank entry (accession number M12486) including the 5′ UTR of *E. coli* fhuA gene[8]; this prediction was subsequently determined to have been confirmed experimentally as corresponding to the *pon* B gene.[5] Other predictions were also made.[1] In view of the high accuracies obtained for correct identification and for rejection of false positives (see above), we expect that these predictions can be verified by the appropriate experimental work.

CONCLUSIONS

In applying various neural net, machine learning, and information theoretic techniques to the problem of detecting protein coding regions in *E. coli* and in Human DNA and to the problem of determining true splice sites from false in Human DNA, neural net methods provided the highest accuracies. These accuracies significantly exceeded those previously reported in the literature. In particular, an accuracy of 99.5% was achieved in determining whether fragments of DNA 180 bases long from *E. coli* were from known coding regions. For DNA from genes known to be expressed in human liver tissue, an accuracy of 98.4% was achieved on 270-base-long fragments. All fragments were ORF's. Discussion of the importance of codon usage and other base correlations was presented. For the problem of distinguishing true splice

---

[1]PREDICTED *E. coli* GENES (GENBANK LOCII, BASE POSITIONS): (ECOAROH1, from 138 to 515); (ECOASPAW, from 1666 to 2763); (ECOCCA, from less than 1 to 386); (ECOCDH, from less than 1 to 292); (ECOEPNMPC, from 481 to 1500); (ECOFHUA, from less than 1 to 320); (ECOFOLA, from less than 1 to 366).

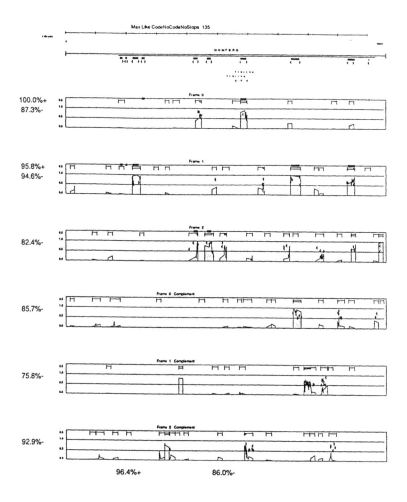

FIGURE 5   Graphical display of results for locus HUMFBRG. The entire figure is divided into different sections. The top section displays scale information and an iconic display of the relevant GenBank FEATURES. Six rectangular boxes follow that contain information both about the output of the neural net and about the sequences themselves. Working from the top down, we first have the title, which reads "Max Like CodeNoCodeNoStops" becase a maximum likelihood neural net method [Eq. (4c)] was used during training on 135-base-long windows within ORF's. Directly under the title is the scale line, which contains tics at 1 kilobase units, as documented at the left of the scale line itself. Under the ends of the scale line are numbers that represent the limits of the GenBank sequence that was analyzed (the entire sequence need not be analyzed). In this case the left limit was 0 and the right 10,564 since those are the boundaries for this GenBank entry. Directly under the scale line is a line that represents the sequence; above this line see HUMFBRG, which is the GenBank LOCUS name for the region that contains the sequence for the region containing (continued)

FIGURE 5 (continued) the human fibrinogen gene. Under the line that represents the sequence there are three other types of icons. One type is a thick solid line, which represents the regions that code for proteins and are annotated as "pept" in GenBank. Additionally, there are right-pointing and left-pointing small triangles that represent splice donor and acceptor sites, respectively. Finally, between these icons and the line representing the sequence, there is a long thin line representing the extent of the RNA transcript. Under these icons, the beginning of the first word(s) from the feature description field in GenBank is displayed—in this case "fibrino" from "fibrinogen gamma chain" and "g·p g" from "g, g-p signal pept."

Each of the six boxes represents the output from the neural networks for different frames: The top three for the same frame as in GenBank, the bottom three for the complementary strand. As the complementary strand is the reverse complement, 5′ to 3′ for transcribed RNA would be right to left instead of left to right.

In the left margin of each box are statistics comparing the output of the net for the Problem 1 coding discrimination to the annotation in GenBank. This is only relevant for ORF's that are bigger than the window size—135 in this case. Additionally, windows are only compared to GenBanks's calls if they are wholly within or outside of a known coding region; overlapping windows are not counted. The statistics are interpreted as follows: If in that frame there are any coding regions containing windows large enough, the percent of those correctly identified as coding by the net are scored as NN% +. Those windows in ORF's that do not code for protein and are correctly identified as not coding, are scored NN% +.

To the left of each box are four numbers. The top number, 0.5, sets the threshold used to determine the minimum signal strength to identify an ORF as coding. The lower three define the positions for lines identifying the maximum signal strength from the net (1.0), the threshold mentioned above (0.5), and the minimum signal strength from the net (0.0).

The most obvious features in each box are shaded regions. The height of this region represents the strength of the signal from the net. Generally, signals above the 0.5 line are identified as coding, and those below are identified as not coding. The window size being finite and not a point source (135 in this case) causes a problem with representation. It is dealt with as follows. At the start of an ORF, the beginning of the window is marked (left on top three boxes, right on bottom three). Subsequently, the end of the window is marked, displaying the entire region presented to the net.

The triangles mark splice sites identified by the net; however, they are only displayed if they are above 0.5 signal strength and in an ORF that has at least one window above 0.4 coding strength. The highest donor and acceptor triangle representations in an ORF are solid; the others are only outlined.

At the top of each box there is additional information. The thick solid lines on the top line represent coding regions, *as annotated in GenBank*, and they are only displayed for the frame in which they would be coding. Directly under them are lines (with ends marked) that denote the extent of the ORF's that were analyzed. For example, in the left part of the figure in frame 0 there are three of these lines without any apparent signal from the net. This occurs because the net signal was so weak that the shaded areas are not visible. (continued)

FIGURE 5 (continued) Sometimes boxes appear under the line that shows the extent of the ORF's. These represent regions identified as coding by the net—the signal from the net was above the threshold.

In this example, most of the windows 135 bases long from coding ORF's were recognized—all of those in frame 0, and most from frame 1. The exons in frame 2 are less than 135 bases, so no positive statistic is generated. There are many windows that are not known to be coding that the net indicates have coding character. For example, see the extreme right-hand end of frame 2 where there is a region that looks like a coding region with a potential acceptor splice site that the net identified as being usable. It is tempting to speculate that there may be a cell type in which alternative splicing causes this region to be expressed.

junctions from false, an accuracy of 91.2% was achieved on intron/exon splice junctions, and 94.5% was achieved on exon/intron splice junctions. The methodology was always to "train" a network on a set of sequences that was distinct from the set used to evaluate the accuracy of the algorithm.

Although these accuracies are cause for optimism about the techniques used in this paper, we caution that much more work needs to be done (in progress) to properly assess their relevance. A single number, the percentage accuracy, is only relevant in the context of the data in which it was obtained, and the proper data to use to derive an index of accuracy is a subject open for discussion. Therefore, we feel that at this preliminary stage equal importance should be attached to general methodology, as opposed to particular percentage accuracies arising from special techniques. In this context various points are made. Simple statistical weighting schemes commonly used in previous methods are related to the simplest neural net approaches. More powerful neural net approaches, therefore, yield additional algorithms that can be used for various problems, and increased accuracies can result. Whether enhanced accuracies are obtained is problem dependent. The practice of testing an algorithm on training data and not on a distinct prediction set (which is not uncommon in the literature) is susceptible to overestimates of accuracy. The notion of a "consensus sequence," constructed entirely from examples of true splice junctions, may be extended to a measure (the Kullback divergence measure) that identifies which base positions contain significant information that allows discrimination of true from false splice junctions.

In summary, initial results were presented from an extensive survey of procaryotic and eucaryotic DNA using various neural net, machine learning algorithms, and information theoretic techniques. High accuracies were obtained, and in so far as comparison to previous techniques are possible at this early stage, we conclude that there is cause for optimism concerning the usefulness of the methods proposed.

ACKNOWLEDGMENTS

This work was done under the auspices of the U.S. Department of Energy and was largely funded by a grant (1-R01-GM40789) from the National Institutes of Health. C. B. was also supported by a grant from the National Institutes of Health (GM-37812). K. S. was additionally supported by a grant from the U.S. Department of Energy (KP-04-04-00-0). R. M. F. and A. S. L. would like to express gratitude for the hospitality of both the Santa Fe Institute (Santa Fe, New Mexico) and the Institute for Theoretical Physics (Santa Barbara, California), where part of this work was performed. We would also like to thank Dr. Dan Davison for a careful reading of the manuscript and for helpful suggestions, and Dr. Patricia Reitemeier for manuscript preparation.

REFERENCES

1. An, Z. "HierTalker." Center for Nonlinear Studies (LANL) Technical Report: LALP 87-27, 1987.
2. Barnes, C., C. Burks, R. Farber, A. Lapedes, and K. Sirotkin. Work in progress.
3. Baum, E., and F. Wilczek. "Supervised Learning of Probability Distributions by Neural Networks." In *Neural Information Processing Systems*, edited by D. Z. Anderson. New York: American Institute of Physics Press, 1988.
4. Broome-Smith, J. K., A. Edelman, S. Yousif, and B. G. Spratt. "The Nucleotide Sequences of the ponA and ponB Genes Encoding Penicillin-Binding Proteins 1A and 1B of *Escherichia coli* K12." *Eur. J. Biochem.* **147** (1985):437-446.
5. Burks, C., J. W. Fickett, W. B. Goad, M. Kanehisa, F. I. Lewitter, W. P. Rindone, C. D. Swindell, C.-S. Tung, and H. S. Bilofsky. "The GenBank Nucleic Acid Sequence Database." *Comp. Applic. Biosci.* **1** (1985):225-233.
6. Cohen, F., R. Arbanel, I. Kuntz, and R. Fletterick. "Secondary Structure Assignment for Alpha/Beta Proteins by a Combinatorial Approach." *Biochem.* **22** (1983):4894-4904.
7. Coulton, J. W., P. Mason, D. R. Cameron, G. Carmel, R. Jean, and H. N. Rode. "Protein Fusions of Beta-Galactosidase to the Ferrichrome-Iron Receptor of *Escherichia coli* K-12." *J. Bacteriol.* **165** (1986):181-192 .
8. DeLisi, C. "Computers in Molecular Biology: Current Applications and Emerging Trends." *Science* **240** (1988):47-52.
9. Denker, J., ed. *Proceedings of the Conference: "Neural Networks for Computing," Snowbird, UT.* New York: American Institute of Physics Press, 1986.
10. Duda, R. O., and P. E. Hart. *Pattern Classification and Scene Analysis.* New York: Wiley InterScience, 1973.

11. Fickett, J. "Recognition of Protein Coding Regions in DNA Sequences." *Nucl. Acids Res.* **10** (1982):5303–5318.
12. Green, M. R. "Pre-mRNA Splicing." *Ann. Rev. Genet.* **20** (1986):671–708.
13. Holland, J. K. Holyoak, R. Nisbett, and P. Thogard. *Induction: Processes of Inference, Learning and Discovery.* Cambridge, MA: MIT Press, 1986.
14. Hopfield, J. J. "Neural Networks and Physical Systems with Emergent Collective Computational Abilities." *PNAS* **79** (1982):2554–2558.
15. Hopfield, J. "Learning Algorithms and Probability Distribution in Feedforward and Feedback Networks." *P.N.A.S.* **84** (1987):8429–8433.
16. The Human Genome Project. "Health and Environmental Saftey Research Committee Advisory Report." Washington, D.C.: U.S. Department of Energy, 1987.
17. Jungck, J. R., and R. M. Freidman. "Mathematical Tools for Molecular Genetics Data: An Annotated Bibliography." *Bull. Math. Biology* **46(4)** (1984):699–744.
18. Kullback, S. *Statistics and Information Theory.* New York: J. Wiley and Sons, 1959.
19. Lachenbruch, P. A., and M. Goldstein. "Discriminant Analysis." *Biometrics* **35** (1979):69-85.
20. Lapedes, A. Paper presented at the Los Alamos Conference on Nonlinearity in Biology and Medicine, 1987.
21. Lapedes, A., and R. Farber. "Genetic Database Analysis with Neural Networks." Abstract. In *Proceedings of IEEE Conference. Neural Information Processing Systems: Natural and Synthetic.* New York: American Institute of Physics Press, 1987, 28.
22. Lee, Y. C. et. al. "Machine Learning Using a Higher Order Correlation Network." *Physica* **22D** (1986):276–306.
23. Lippmann, R. "Acoustics, Speech, and Signal Processing: An Introduction to Computing with Neural Nets." *IEEE ASSP Magazine* **4** (1987):4–22.
24. Mount, S. "A Catalogue of Splice Junction Sequences." *Nuc. Acids Res.* **10** (1982):459–472.
25. Nakata, K., M. Kaneisha, and C. DeLisi. "Prediction of Splice Junctions in mRNA Sequences." *Nuc. Acids Res.* **13** (1985):5327–5340.
26. Omohundro, S. "Efficient Algorithms with Neural Network Behavior." *Complex Systems* **1** (1987):273–347.
27. Polack, S., H. Hicks, and W. Harrison. *Decision Tables.* New York: Wiley-Interscience, 1971.
28. Press, W., B. Flannery, S. Teukolsky, and W. Vetterling. *Numerical Recipes: The Art of Scientific Computing.* New York: Cambridge University Press, 1986.
29. Quinlan, J. "Induction of Decision Trees." *Machine Learning* **1** (1986):81–106.
30. Rummelhart, D., and J. McClelland. *Parallel Distributed Processing.* Vol. 1. Cambridge, MA: MIT Press, 1986.

31. Sejnowski, T. J., and C. R. Rosenberg. "Parallel Networks that Learn to Pronounce English Text." *Complex Systems* **1(1)** (1987):145–168.

32. Sellers, P. H. "Pattern Recognition in Genetic Sequences." *PNAS* **76** (1979):304.

33. Shapiro, M. B., and P. Senapathy. "RNA Splice Junctions of Different Classes of Eukaryotes: Sequence Statistics and Functional Implications in Gene Expression." *Nuc. Acids Res.* **15** (1987):7155–7174.

34. Staden, R. "Computer Methods to Locate Signals in Nucleic Acid Sequences." *Nucl. Acids Res.* **12** (1984):505–519.

35. Stormo, G. D., T. D. Schneider, L. Gold, and A. Ehrenfeucht. "Use of the 'Perceptron' Algorithm to Distinguish Translational Initiation Sites in *E. coli.*" *Nuc. Acids Res.* **10** (1972):2997–3011.

36. Trifinov, E. "Translation Framing Code and Frame-Monitoring Mechanism as Suggested by the Analysis of mRNA and 16 S rRNA Nucleotide Sequences." *J. Molec. Biol.* **194** (1987):643–652.

37. Waterman, D. *A Guide to Expert Systems.* New York: Addison-Wesley, 1978.

M. N. Liebman and A. L. Brugge
BioInformation Section, Biotechnology Division, Mail Code F-2, Amoco Technology Company, Naperville, IL 60540

CPGA[1]: A Prototype for Analysis of the Sequence-Structure-Function Relationship in the Human Genome Project

INTRODUCTION

From a long-term perspective, the development of an interface between computational science and nucleic acid sequencing is a significant but partial solution to the complex computational and biological issues raised within the Human Genome Initiative (HGI). While mapping and sequencing the genome, the concomitant problems of database generation and interfacing seem to be the major issues facing HGI-researchers today; these areas appear to be bound by the limits of present technology, i.e., solutions do exist but they are expensive, tedious and inefficient. The immediacy, visibility and magnitude of sequencing and mapping frequently overshadow the consideration that these are but the first of several major components of the HGI. Two other major areas involve use of sequence data to predict the three-dimensional structure of the constituent proteins and nucleic acids, and, more significantly, to predict their biophysical function, *in vitro* and *in vivo*. The potential for development of these latter capabilities is presently limited by our scientific understanding alone rather than by technology.

[1]Center for Prokaryote Genome Analysis.

Our research program concentrates on the computational aspects of each of these areas through: (a) in-house research efforts; (b) external research and database collaborations, e.g., Fourier-transform infrared database with the U.S. Department of Agriculture; and (c) funding of external collaborations, e.g., a research grant for development of sequence analysis software at the Center for Prokaryote Genome Analysis (CPGA) at the University of Illinois, Champaign-Urbana. In this report, we describe:

A. our approach to integrate the research areas outlined above;

B. the model systems we use as test applications of the representational and analytical methodologies;

C. an overview of some of our methods for data representation, i.e., structure, function and properties, which support the integration of this information; and

D. examples of both the methodological development and the application to a specific protein family.

BACKGROUND

Science is the observation, identification, description, experimental investigation and theoretical explanation of natural phenomena. This definition extends to several of the disciplines within the Human Genome Initiative including genetics, evolution, biology, chemistry, physics, mathematics and computer science. The individual evolution of these disciplines as highly specialized, technologically oriented domains of knowledge has limited communication amongst practitioners, both inter- and intra-discipline. This inefficiency in communication results in the generation of large, quantitative, domain-specific databases which may be limited in their accessibility and/or usefulness to other scientists. In the context of our research, we consider present limitations in database access to extend beyond telecommunications to database architecture and in database usefulness to extend to evaluation of the data included.

By means of example we cite the existing lack of correlation among three primary databases, GenBank,[12] Protein Identification Resource (PIR)[32] and Protein Data Bank (PDB),[2] and use illustrations from PDB. The obvious links, involving the parallel occurrence of observed or derived amino acid sequences or by enzyme/protein classification identifier, are only now being activated. *Functionally oriented descriptors*, for example, enzyme family, catalytic mechanism, and *structurally oriented* descriptors remain unsearchable across PDB and without correlated descriptors in GenBank and PIR. In addition, these descriptors are provided by the depositor and may lack the uniformity in definition that is essential for correlation and analysis. Examples of such subjective descriptors in PDB include secondary structure definitions and boundaries, amino acid sequence (i.e., the "sequence" record and sequence in "atom" records may differ because of physiological processing or because of limitations in experimental observation, without notation) and special sites including catalytic, binding site definitions. The inability to identify the existing structures and sequences which are serine proteases serves as an example of the limitations of database linkage which presently hinder researcher queries.

A. A MODEL FOR THE BIOINFORMATION PATHWAY

We present, in Figure 1, a bioinformatic perspective of the research areas within the HGI, weighted to reflect our research emphasis on potential uses for the genomic sequence. The axes for this plot may be viewed as *data intensive* along the ordinate and *relationally intensive* along the abscissa. Thus the domain-specific databases appear in regions perpendicular to the x-axis: *nucleic acid sequence databases* (GenBank, EMBL and Genome Projects such as *E. coli* and CPGA); *amino acid sequence databases* (PIR and Swiss Protein Sequence); *structural databases* (PDB protein and nucleic acid structures including x-ray and NMR structures); *physicochemical property databases* (circular dichroism,[28] FT-IR,[5,33] nuclear magnetic resonance,[29] physical properties,[4] *in vitro* specificity[4] and molecular dynamics trajectories); and *physiological function databases* (*in vivo* specificity and reactivity,[4] and enzyme cascades). The links between these databases and the rules that span the sequence-structure-function relationship extend parallel to the x-axis. That knowledge which is currently accessible would appear primarily as a series of

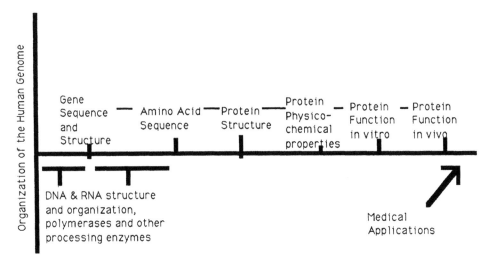

FIGURE 1   A model of the bioinformational pathway of the human genome.

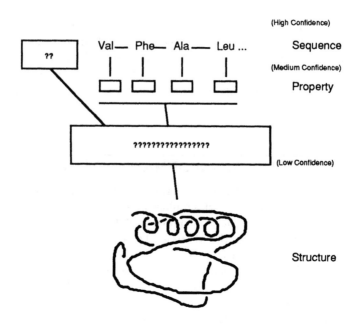

FIGURE 2   Relative stages and confidence levels of protein structure prediction from amino acid sequence and ???.

vertical lines whose distribution would be skewed towards the origin (sequence end) of the x-axis.

This observed bias of existing, organized knowledge results from the technological advances in sequence determination, and the limited progress in identifying rules that govern the relationship among sequence, structure and function (SSF). We can characterize those relationships that have been noted in terms of their reliability (Figure 2). Thus such rules would include *high-confidence rules:* translation of DNA coding regions into resulting amino acid sequence; *medium-confidence rules:* prediction of limited secondary structural features from amino acid sequences and/or properties, prediction of the three-dimensional structure of homologous proteins, correlation of protein structure with spectroscopic data; and *low-confidence rules:* prediction of protein tertiary structure from amino acid sequence data, understanding of the structure-function relationship in proteins including those of known structure.

B. ANALYSIS OF SSF: THE "CPGA" AND "SP" APPROACHES

The long-term goal of our research is to identify fundamental relationships which can span the data from gene sequence to physiological function. To achieve this, we concentrate on two distinct but related approaches: (1) data acquisition of the genomic sequence, physical map and related biological matrix for a specific organism, and (2) data integration and analysis for a single family of proteins where available information spans the range of bioinformation presented in Figure 1.

The former approach, *genomic analysis of a specific organism*, is typified by our involvement with CPGA and our interest in other genome sequencing efforts to serve as prototypes for evaluating the requirements, limitations and potential for success of HGI. The integration of data will proceed from sequence determination, physical mapping, regulatory and coding regions, relationship of encoded metabolic pathways, etc. This is representative of the problem of the mini-biological matrix, but one for which the data will be available more rapidly, and provide simpler, more readily testable areas for analysis and prediction. A detailed description of the sequencing targets and efforts of CPGA are presented elsewhere.

The latter approach, *structure-function analysis of an evolutionary family of proteins*, attempts to identify the SSF relationship among the serine proteases (SP), by comparison and analysis of the gene organization, amino acid sequences, three-dimensional structures, physico-chemical properties, specificities and reactivities *in vitro* and *in vivo* of both eukaryotic and prokaryotic proteases, including those apparently linked by convergent evolution of mechanism, alone. We utilize this approach to distinguish among: (a) the *inadequacy of present experimental observations* to describe the SSF link; (b) the *inadequacy of present computational tools* for analysis and integration of theoretical and experimental data; and (c) the potential that *no single set of rules will enable prediction* of the SSF relationship. Although the hypothesis that no such predictive rules exist receives little attention or acceptance, we must acknowledge the inherent limitations of the data upon which we assert this bias and determine the validity of this assumption.

Our application of the methodologies outlined in Figure 1 and described below towards the serine proteases is undertaken because of the length of the "chain-of-information" that exists for this family of proteins.[1,9,10,11,14,20,25,30]

1. Physiological Significance: Enzymes in this family are involved in physiological processes that range from simple peptide digestion to highly organized physiological control networks such as coagulation and complement activation and to disease processes such as extracellular matrix protein degradation, inflammation and arthritis, emphysema and tumor cell metastasis. This family of enzymes has been the subject of study by more investigators and with more experimental methodologies than any protein system other than hemoglobin or immunoglobulin. The hemoglobin all function in oxygen transport, with variation in sequence impacting on the physiological parameters rather than alternative pathways. Immunoglobulins function by sequence modifications impacting on antigen specificity, but with that recognition involved within a single

systemic function. By contrast, service protases are involved in a broader range of biologically important systems where the observed evolutionary changes in amino acid sequence yield more divergent functional differences (i.e., specificity and reactivity).[30]

2. Genetic information: These enzymes are essential components of both eukaryotic and prokaryotic systems. The full sequence of the encoding gene is known for several of these proteins, including several of the coagulation enzymes, e.g., thrombin, as well as trypsins, chymotrypsins and elastases from several organisms, and prokaryote genes for both trypsin-like and subtilisin-related serine proteases.[1,3,9,10,12]

3. Amino Acid Sequence: Amino acid sequence data is available for far more proteases than are the corresponding DNA sequences. These span multiple physiological processes within a single organism, equivalent processes amongst different organisms, genetic variants within an organism and both eukaryotic and prokaryotic systems.[11,14,32]

4. Three-Dimensional Protein Structure (X-ray, NMR): The three-dimensional structures of a number of serine proteases have been determined to high resolution by X-ray crystallographic methods and are available within the Protein Data Bank (PDB). These include examples from both eukaryotes (e.g., bovine, porcine, rat) and prokaryotes (e.g., myxobacter, *streptomyces griseus, b. subtilis, tritirachium album*). In addition, several of these enzymes have been examined in a variety of conformational states, at various pH's, complexed with natural and synthetic inhibitors, as zymogens, etc.

5. Physico-Chemical Properties: These enzymes have been the target of extensive *in vitro* studies, probing conformational changes which accompany inhibitor/substrate complex formation, zymogen activation, perturbation of solution environment, etc. Such data include spectroscopic measurements, e.g., nuclear magnetic resonance (NMR), Fourier-transform infrared (FT-IR), optical rotatory dispersion (ORD), circular dichroism (CD), fluorescence and ultra-violet (UV) spectroscopy, as well as sedimentation equilibrium studies of aggregation and molecular recognition properties.

6. Protein Function, *in vitro:* The specificities directed towards synthetic substrates, inhibitors and transition-state analogues have been studied for all examples of the serine proteases because of the need to modulate their activity within specific normal physiological and disease states. This has been an intensive example of drug design. More recently these enzymes have become the subject for research within biotechnology as they are potential targets for gene-replacement therapy, engineered-enzyme replacement therapy and inhibitor-design, and in areas which range from medicine to agriculture.

7. Protein Function, *in vivo:* This family of enzymes is found in both eukaryotes and prokaryotes, in a variety of tissues, membrane-bound and circulating and each exhibiting physiological specificity which can be correlated with structural and sequence-based differences. The occurrence of these enzymes in physiological cascades in which homologous enzymes are involved in zymogen activation,

enzyme inactivation, inhibitor and substrate interaction, suggest that an even higher-level of evolution-based functional similarity may exist.[30]

8. General Applicability of Approach (Technology): Developed for this study, it should be generalizable to other protein/enzyme families for sequence-structure-function analysis. Information can be related, through the design and implementation of appropriate knowledge acquisition/database software, to capabilities/limitations of specific physical measurements, definitions of structural classes, site of action (i.e., tissue or physiological environment), etc., for use in "rule" generation.[13]

9. Site-Directed Mutagenesis: Several members of this enzyme family, e.g., subtilisin, trypsin, and alpha lytic protease, have been extensively probed by site-directed mutagenesis for perturbation of reactivity, specificity and enzyme mechanism. In addition, high-resolution three-dimensional structures are available for many of the bacterial trypsin-like and subtilisin protease mutants.[3,6,9,10]

RELATIONAL IMPLEMENTATION OF AN EXTENDED-PROTEIN DATA BANK

With the extensive efforts to implement relational database formats for the nucleic acid and amino acid sequence databases, and our interest in extending the integration of sequence with structure and function data, we have undertaken an extension of the existing Protein Data Bank. This restructuring is more than conversion to a relational format, or linking of sequence information between structure and sequence databases as we outline below in several target areas.

At present PDB contains several data/comment fields that are supplied by the structure depositor. Several groups are attempting to make this information available within a more readily accessible format, a relational database, which would allow for both horizontal transversing, through a specific protein file, as well as vertical, i.e., within a specific record type across different proteins. This is somewhat limited by the nature of the present data in PDB, which is not generated in an analytic manner for each protein, thus limiting its true relational potential. In addition, information of potential significance for studying the structure-function relationship may not be presently expressed in the databank, although derivable or collectable from secondary sources. We have identified several systematically applied (and relationally addressable) descriptors within classes such as:

- Structure Determination: structural resolution; structural refinement method; method of data collection; crystallization conditions, e.g., solvent, pH, temperature; presence of non-crystallographic intra- and inter-molecular symmetry.
- Structural Parameters: folding domains; sequence of secondary structure (substructure); substructure composition; cis-peptides; deviations from ideal configuration in distances or angles.
- Physico-Chemical Parameters: molecular volume, solvent accessible surface; rugosity (surface area to volume ratio); ellipsoid of revolution parameters, e.g.,

axial ratio through center of mass, orientation; radius of gyration; molecular dipole moment, including backbone and side-chains; packing density; bulk hydrophobicity; amphiphilicity; molecular weight; link to spectroscopic data bases.

■ Function *in vitro*: substrate specificity; inhibitor specificity; binding pocket based on substrate/inhibitor contacts; mechanism; binding constant/turnover rate; antigenic determinants.

■ Function *in vivo*: phylogenetic source of protein; organ site of activity; environmental conditions at site of activity; substrate specificity; inhibitor specificity; co-factors; binding constant/turnover rate; participation in metabolic pathway or cascade; link to nucleic acid database sequences, e.g., exon boundaries; Enzyme Commission classification; membrane localization; post-translational modification, e.g., glycosylation, sites and type; isolatable form, e.g., zymogen, preproenzyme, proenzyme.

This list of descriptors is not considered to be complete but rather indicative of the information that we are presently linking to existing and newly added proteins in PDB.[12,32] It is additionally the intent of this relational approach to enable inheritance of characteristics and descriptors by type, e.g., a structural element has physico-chemical descriptors, a protein may have multiple structural elements (domains or substructures) each of which will have its own physico-chemical descriptors as well as those that represent the intact protein. This relational construct to extend the knowledge base of the PDB, as well as to establish closer linkages to other significant databases, including GenBank, EMBL, and PIR, has been initiated in collaboration with the Center for Prokaryote Genome Analysis at the University of Illinois, Champaign-Urbana, funded in part through a grant from Amoco Technology Company's BioInformation Section of the Division of Biotechnology.

C. REPRESENTATION OF INFORMATION FOR ANALYSIS AND COMPARISON

To integrate data and information from the various experimental and computational areas described above is dependent upon the ability to: (1) represent the data in a form that minimizes biases; and (2) optimize access to structural, functional and physico-chemical data through the use of common algorithms.

We describe our methods for representation, analysis and comparison[17,18,19,20,21,22,25,35] in terms of their dimensionality: one-dimensional representations relate descriptors to a relative position within the nucleic acid/amino acid sequence, using the sequence to derive the list or ordering; two-dimensional representations also use a sequence-based ordering to generate a matrix containing descriptors which relate pairs of sequence elements, i.e., residues, which may not

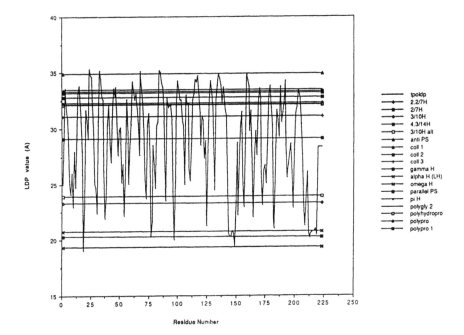

FIGURE 3   Linear distance plot of trypsin showing standard conformations.

be contiguous three-dimensional representations relate descriptors to positions relative to a defined coordinate system, typically Cartesian or polar coordinate space; and four-dimensional representations involve descriptors which incorporate time with lower-dimensional descriptors. Structural, functional and physico-chemical descriptors using each of these formats are described below, along with their respective advantages and limitations. *An important benefit results from the regions of overlap or redundancy that exist amongst these methods of representation as it enhances the opportunity to make observations and to validate their consistency.*

One-dimensional representations use the position within a sequence to represent a descriptor based on that specific sequence element, or evaluated within a window of sequence elements which bounds that specific sequence element. With this representation we use descriptors for:

1. Structure: Linear distance plot (LDP, LDP= sum $\{d(ij), j = i+1 \text{ to } i+4\}$ for each successive alpha carbon, with sum for residue i plotted versus sequence position i) (Figures 3–5); bond dihedral angle plot (BDA).[18,19,20,37]
2. Physico-Chemical Properties: as derived from statistical analysis of observed protein/nucleic acid structures, and including hydropathy, flexibility, bulk, surface area, and pK.[37]
3. Functional Properties: relative dipole orientation of sequence elements.[27]

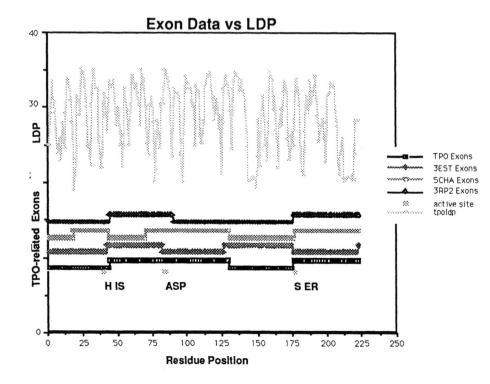

FIGURE 4   Comparison of serine protease exon boundaries with LDP, and active site residues.

Advantages realized with one-dimensional representations include their use in rapid pattern recognition and analysis of structure, function and properties, separately and in combination; invariance of a given mapping to rotation and translation of the reference frame of the descriptor, e.g., rotation or translation of a molecule does not alter its linear distance plot; extendability to other descriptors as developed, e.g., intron-exon boundaries (Figure 4), enzyme-inhibitor contact regions; and suitability for direct comparison of localized regions between closely related molecules, e.g., local conformational perturbation upon inhibitor binding into an enzyme.[20,21] The main limitations involve the chiral ambiguity in structural representation of a protein with the LDP, i.e., distance information within a local structure lacks chirality, and selection of "window" size to preserve information content when representing window-based descriptors, e.g., averaging hydropathy over a variable window size of residues and representing a single value for a specific sequence position (see also Figure 5). Analysis and comparison tools

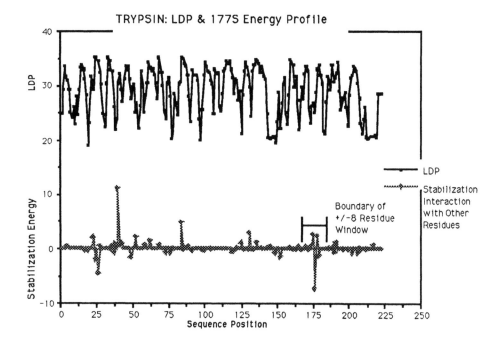

FIGURE 5 Limitation of window size on structural and energetic perspective.

that have proved most useful with these representations include use of signal processing methods such as simple addition or subtraction to highlight local regions among closely related molecules, and dynamic programming algorithms for pattern recognition that encompasses potential insertions and deletions and has been used to develop a protein substructure library.

Two-dimensional representations are based on the generation of an $n \times n$ matrix, where n is the number of elements within the sequence list. With this representation format we use descriptors for:

1. Structure: distance matrix, partitioned distance matrix, both intra- and inter-molecular[17,20,21] (see Figure 6, which contrasts distance matrix with information content expressed by viewing ±8 residue window about each residue position which would be a one-dimensional representation).
2. Physico-Chemical Properties: weighted hydrophobicity, base-pairing potential.[23,25]
3. Function: electrostatic potential matrices for backbone, side-chains including dipole-dipole, charge-dipole and charge-charge interactions, dipole-interaction matrix.[23,24,25]

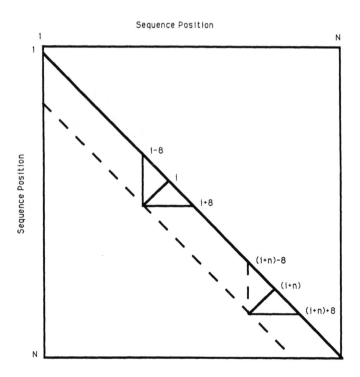

FIGURE 6 Limitation of information concerning tertiary structure imposed by boundary set of ±8 residues.

This representation has advantages in analysis by pattern recognition of structural organization and orientation and comparison of independent molecules or conformational states of a specific molecule. The natural organization of this transform yields sequence-adjacent features, i.e., secondary structure-related, near to the matrix diagonal, and more sequence-distant features more distant from the matrix diagonal. As with the one-dimensional representation, these representations are invariant to the orientation of the reference frame for these descriptors. Where the one-dimensional representation of small, individual structural segments is identical for mirror-image structures, the two-dimensional representation of the intact molecule benefits from the tendancy of nature to not present both a protein and its mirror image. This ambiguity for mirror structures does persist, however, in the analogous two-dimensional representation of small molecules which can occur as mirror images in nature. Comparison and analysis utilize pattern recognition and vision-analysis based tools, and are presently being evaluated within a neural network training project.

Three-dimensional representations involve the direct referencing of the structure-function-property descriptors to an external (cartesian) or internal coordinate (polar) reference frame. These representations comprise the more typical manner for viewing the three-dimensional structure with interactive, three-dimensional computer graphics displays. The descriptors which we use in this form include:

1. Structure: atomic coordinates (orthogonal reference frame), fractional atomic coordinates (crystallographic reference frame), polar plot around designated atom/point centers, molecular graphics representation.[20,21]
2. Physico-Chemical Properties: solvent accessible surface, van der Waals surface, molecular dipole orientation.
3. Function: hydrodynamic shape, molecular electrostatic potential surface, ion/ solvent channels, internal cavities.[19,26]

The advantages of this form of representation are in its wide-spread acceptance due to the increased availability of interactive computer graphics displays. This advantage, which has evolved to make these representations readily accepted by the non-specialist scientific community, may prove to be a long-term limitation because of its interpretation as representing "what a molecule looks like." The most notable limitation of this form of representation is the dependence of a particular viewing perspective on the orientation within a coordinate reference frame. Thus it is necessary to rotate a molecule, i.e., apply a rotation matrix to the atomic coordinate list and re-display, to obtain a different view. This reveals the difficulty in applying analytical methods to the analysis of these representations. Comparison and analysis are performed by effecting the direct superposition of one molecule onto another by generation of a rotation-translation matrix, orientation sampling within the display reference frame, e.g., rotating the molecule in 5-degree intervals about a specific viewing axis, and visual comparison.

Four-dimensional representations involve those descriptors which contain an element of time in conjunction with other relevant descriptors. Thus any of the descriptors described above could be utilized in a four-dimensional representation due to fluctuation of structure, function or physico-chemical properties with time. Those descriptors which we identify to have specific temporal components are:

1. Structure: dynamical nature of a molecule as viewed by computation of a molecular dynamics trajectory; structural disorder observed in crystallographically determined structures; structural dynamics as viewed by NMR.
2. Physico-Chemical Properties: measurable properties which change with time, principally due to the occurrence of functionally derived processes.
3. Function: enzyme cascades, metabolic pathways, allosteric processes as mapped through sparse matrix methods.[27,30]

The advantage of this form of representation is the ability to incorporate processes which are the true physico-chemical and functional states of the molecules under study. The dynamic nature of macromolecules may be closely linked with functions of reactivity and specificity. Analysis and comparison of pathways that contain molecules of common evolutionary origin may reveal the extent to which evolutionary influence is exerted at the physiological level and lead to the under-standing of structure-function at a higher level of development, and a much fuller appreciation of the information actually encoded at the genetic level.[7]

D. APPLICATIONS OF THE METHODS FOR REPRESENTATION AND ANALYSIS

As we have noted in this report, the full synthesis of bioinformation requires utilizing multiple representations of structure, physico-chemical properties and function as presented above, both singly and in combination. To illustrate such processes, two specific examples are described in terms of the questions which were addressed and the manner in which these approaches have evolved.

1. ELECTROSTATIC ENERGY ANALYSIS OF PROTEIN STRUCTURE[22,23,24]

This method (see Figure 5) involves the representation and analysis of the organi-zation of the electrostatic energy components present in the tertiary structure of a protein for use in the study of conformational pathways, evolutionary constraints derived from function, structural basis for allosterism, physical environment of in-dividual residues, and predicting potential effects of site-directed mutagenesis.

The application of this method involves the deconvolution of a protein struc-ture into its constituent peptide and side-chain dipoles, using vector addition of the individual bond moments assigned to the observed side-chain conformation, i.e., glu-tamine side-chain dipole ranges from 3.03 to 4.11 debyes in native trypsin (1TPO) depending on the observed conformation of the individual glutamine. The peptide dipole computed in this manner for a trans-peptide is $3.81 \pm .02$ debye on average, and 2.71 debyes ± 0.02 for a proline, with cis-peptides ranging from 2.2–2.4 debyes. These observations are based on calculations performed on high-resolution protein structures, and all proteins are rigorously screened for a wide-range of factors which would limit their use in more rigorous analysis such as this computation.

Potential energy computations are carried out by assigning formal charges based on individual amino acid pK's, using a uniform dielectric constant of $\epsilon = 2$ in the initial studies, and using standard forms of the potential functions for computation of dipole-dipole, charge-dipole and charge-charge interactions. Using the dipoles computed from the observed structures as contrasted with an average value assigned to the 20 amino acids, introduces the details of the actual crystallographic structure. Additionally, separate computations of the interactions, side chain:side chain, side chain:main chain and main chain:main chain, enable the further separation of those terms which may result from backbone conformation, e.g., similarities amongst homologous proteins, from those that reflect variation in amino acid sequence, either through natural evolution or synthetic evolution, i.e., site-directed mutagenesis.

While this approach is primitive in its analysis of higher-order components of the electrostatic energy, many of these enhancements, e.g., polarizability, variation in local dielectric constant, solvent interaction, can be readily accommodated within the framework of the computation. More significant than these limitations are the analytical potential presented within the one- and two-dimensional representations described above. As the computation is performed on a residue-residue basis, the two-dimensional matrix output form permits the generation of a component of the energy directly correlated with a representation of the three-dimensional structure. In this manner the analysis is immediately accessible to the analysis tools used to describe the structural organization of the protein, now in terms of the stabilization or destabilization energies within local secondary structure, tertiary structure, structural domains, between subunits, etc. The linkages between a given amino acid and the remainder of the protein are also highlighted in a manner which enables assessment of the potential effects of specific site-directed mutations (SDM). The obvious extension of this representation to one-dimension has enabled the comparison of profiles of the environment of specific amino acids within the same protein, e.g., environments of different tryptophans in trypsin, as well as profiles of the equivalent residue in evolutionarily related proteins, e.g., SER 195 equivalent among various serine proteases (Figure 5).

This analysis is being used to contrast sequence homology with functional variability, determine the range of structural features, coupled with specific sequences, which can generate analogous physico-chemical properties with variation in conformation, relationship between observed structural perturbation and energetic perturbation, energetic sources of enzyme specificity in limited proteolysis and to identify targets for SDM distant from the active site, to modulate specificity and reactivity as expressed within the active site.

2. FOURIER-TRANSFORM INFRARED SPECTROSCOPY AND A SUBSTRUCTURE LIBRARY[5,19,20,32]

The ability to accurately correlate a spectroscopic technique (see Figure 3) with high-resolution protein structure assignments has long been the goal of experimental biophysics. Until recently neither the experimental methods nor the available structural information and algorithms for their analysis were suitable to achieve this goal. Our integration of both approaches has revealed significant success with refinements in progress. The method of Fourier-transform infrared spectroscopy has evolved to present a signal-to-noise ratio of approximately 10,000 to 1, a resolution of 2 cm-1 and facilities to measure 4000 spectra/30 minutes. Additionally, we concentrate on the spectral region of the Amide I, II and III bands, which are primarily associated with peptide linkages rather than side-chain interactions. The observation that each resolved band is associable with a unique set of resonances, leaves the singular problem of identifying and assigning the structural feature with the correct bands.

We have developed two parallel approaches to this problem, both of which utilize the generation of a substructure library from the observed protein structures, based on the linear distance plot and a pattern recognition implementation using dynamic programming. Where previous structural analyses have typically applied 3- or 4-state analysis of known protein structures, e.g., helix, sheet, turn and coil, the predictive accuracy of these descriptors is limited to approximately 60–65%. This has been the result of classification bias, i.e., assuming that all structural features could be *adequately* assigned to these classifiers. Our analysis has shown that proteins can be more completely described by a library of substructures, i.e., contiguous conformational features, of 8-residue lengths, which span approximately 98–99% of all known protein structures. An important criteria is that assignment of a substructure requires observation in more than one protein. This library subclassifies previous structural classifiers and reproduces these results using these coarse classifiers.

Accurate assignment of these substructures to observed spectral bands is being carried out in two parallel approaches:

a. Our previously observed conformational perturbation of trypsin by inhibitors, from crystallographic analysis is being analyzed in terms of substructure perturbation and experimental measurements have been carried out on the equivalent protein states. Difference FT-IR spectra will permit assignment of perturbed bands to perturbed substructures. This approach will be recursively applied to the homologous serine proteases, and then to other proteins which contain the assigned substructures to refine the assignments.

b. Spectra have been determined for a wide range of proteins for which structure-substructure information is available from crystallography. We are attempting to use unsupervised learning algorithms based on neural network analysis methods to effect the initial assignments of spectral bands to substructures which will also be recursively evaluated by extension to other proteins in the training set.

It should be noted that the generation of the substructure library for this study is directly applicable to examine sequence relationships among the equivalent substructures, as well as other descriptors as discussed above. The extension of these results can further the ability to experimentally observe conformational linkages in proteins. This is accomplishable by performing SDM or other specific modifications or intermolecular complexation, and measuring the FT-IR spectra. The substructure assignments will enable the determination of which structural features are being conformationally perturbed independent of their spatial relation to the site of chemical modification in the protein. Additionally, analysis of proteins of unknown three-dimensional structure should permit identification of component substructures and generation of predictions of tertiary structure for further evaluation.

FUTURE DIRECTIONS

Much of our present and future research involves the development of supporting resources for the programs outlined in this report, both through internal research programs and external collaborative efforts. A major thrust is in the exploration of machine learning approaches to the study of the sequence-structure-property-function relationship. Major components of this effort are collaborations with K. Spackman of Oregon Health Sciences Center and G. Wilox and M. Poliac of the Minnesota Supercomputer Institute. Our underlying philosophy and emphasis center about developing an understanding of this complex relationship without concentrating solely on the ability to predict structure from sequence which may require additional information (Figure 2). Within this relationship are the key elements necessary for structure prediction, and the ability to predict structure without an understanding of how it works or how to modulate its function would leave us far too short of the goals outlined in Figure 1.

SUMMARY

The resources and efforts being devoted to the Human Genome Initiative and its prototype genome projects should not lose sight of the long-term goals beyond sequencing and mapping. We view HGI as a large, broad peak which we can deconvolute into three gaussians: sequencing and mapping; structure prediction; and structure-function analysis. We recognize that other researchers may resolve additional hills. At present most efforts are targeted towards the "first hill," but it should be remembered that upon conquering that, two more remain which at present appear to be of even greater height. The integration of bioinformation, through computational representations, some of which are presented in this report, attempts to prevent these hills from drifting too far apart. The success of this integration will enable some of the momentum generated to climb each hill to be used to mount the next. This potential advantage should not be lost but fostered through interactive and collaborative multidisciplinary research efforts involving both the public and private sectors.

ACKNOWLEDGMENTS

It is most appropriate to acknowledge the students, postdoctorals and collaborators who have contributed over the years to various aspects of the research outlined in this report. The author wishes to acknowledge the contributions of S. V. Amato, R. Buono, R. Lanzara, S. Prestrelski, R. Ting, Dr. A. L. Williams, Jr. , Dr. T. F.

Kumosinski, Dr. M. Byler, Dr. J. C. Lee, Dr. A. Lipkus and the late Dr. H. Susi. In addition we acknowledge resources and facilities provided at the Institute for Cancer Research; Departments of Pharmacology and Physiology/Biophysics, Mount Sinai School of Medicine; Battelle Laboratories (Columbus); and U.S. Department of Agriculture, Agricultural Research Service, Eastern Regional Research Laboratory.

REFERENCES

1. Bell, G.I., C. Quinto, M. Quiroga, P. Valenzuela, C.S. Craik, and W. J. Rutter. "Structure of the Rat Pancreatic Chymotrypsin B Gene." *J. Biol. Chem.* **259** (1984):14265–14270.

2. Bernstein, F. C., T. F . Koetzle, G. J. B. Williams, E. F. Meyer, Jr., M. D. Brice, J. R. Rodgers, O. Kennard, T. Shimanouch, and M. Tasumi. "The Protein Data Bank: A Computer-Based Archival File for Macromolecular Structures." *J. Mol. Biol.* **112** (1977):535–542.

3. Bone, R., J. L. Silen, and D. A. Agard. "Structural Plasticity Broadens the Specificity of an Engineered Protease." *Nature* **339** (1989):191–1955.

4. Brugge, A. L., and M. N. Liebman, unpublished.

5. Byler, D. M., and H. Susi. "Examination of the Secondary Structure of Proteins by Deconvolved FT-IR Spectra." *Biopol.* **25** (1986): 469–487.

6. Carter, P., and J. A. Wells. "Engineering Enzyme Specificity by 'Substate-Assisted Catalysis.'" *Science* **237** (1989):394-399.

7. Conselor, T., M. N. Liebman, and J. C. Lee. "Domain Interaction in Rabbit Muscle Pyruvate Kinase: Small-Angle Neutron Scattering and Computer Simulation." *J. Biol. Chem.* **263** (1988):2794–2801.

8. Conselor, T., M. N. Liebman, and J. C. Lee. "Domain Interaction in Rabbit Muscle Pyruvate Kinase: Intersubunit Contacts and the Allosteric Switch." Submitted for publication to *Biochemistry.*

9. Craik, C. S., T. Fletcher, S. Roczniak, P. J. Barr, R. Fletterick, and W. J. Rutter. "Redesigning Trypsin: Alteration of Substrate Specificity." *Science* **228** (1985):291–297.

10. Craik, C. S., S. Roczniak, C. Largman, and W. J. Rutter. "The Catalytic Role of the Active Site A Spartic Acid in Serine Proteases." *Science* **237** (1987):909–913.

11. DeHaen, C., H. Neurath, and D. C. Teller. "The Phylogeny of Trypsin-Related Serine Protease and Their Zymogens: New Methods for the Investigation of Distant Evolutionary Relationships." *J. Mol. Biol.* **92** (1975):225–259.

12. GenBank, A national resource for nucleic acid sequences, maintained by a contract from NIH-DRR to Los Alamos National Laboratory and Intelligenetics, Inc., Mountain View, CA.

13. Greer, J. "Model for Haptoglobulin Heavy Chain Based on Structural Homology." *Proc. Natl. Acad. Sci. USA* **77** (1980):3393–3397.

14. Hartley, B. S. "Homologies in Serine Proteases." *Phil. Trans. R. Soc. London Ser. B* **257** (1970):77–87.
15. Kabsch, W., and C. Sander. "Dictionary of Protein Secondary Structure: Pattern Recognition of Hydrogen-Bonded and Geometrical Features." *Biopol.* **22** (1983):2577–2637.
16. Levitt, M., and J. Greer. "Automatic Identification of Secondary Structure in Proteins." *J. Mol. Bio.* **114** (1977):181–293.
17. Liebman, M. N. "Quantitative Analysis of Structural Domains in Proteins." *Biophys. J.* **32** (1980):213–217.
18. Liebman, M. N. "Topographical Analysis of Specificity in Chemotherapeutic Systems." *Prog. Clin. Biol. Res.* **172B** (1985):285–299.
19. Liebman, M. N., C. A. Venanzi, and H. Weinstein. "Structural Analysis of Carboxypeptidase A and Its Complexes with Inhibitors as a Basis for Modeling Enzyme Specificity." *Biopol.* **24** (1985):1721–1758.
20. Liebman, M. N. "Distance Approaches to Protein Structure Analysis and Prediction." *J. Cell. Biochem.* **9B** (1986a):132.
21. Liebman, M. N. "Structural Organization in the Serine Proteases." *Enzyme* **36** (1986b):115–140.
22. Liebman, M. N. "Molecular Modeling of Protein Structure and Function:A Bioinformatic Approach." *J. Computer-Aided Molec. Des.* **1** (1987):323-341.
23. Liebman, M. N., and T. F. Kumosinski. "Analysis of the Structure-Function Relationship in Proteins using Two-Dimensional Energy Profiles." *Biophys. J.* **51** (1987):450a.
24. Liebman, M. N., and F. G. Prendergast. "Electrostatic Interactions of the Tryptophan Residue in Streptomyces Griseus Proteinase A and Ribonuclease T1." *Biophys. J.* **51** (1987):2761.
25. Liebman, M. N. "Analysis of the Biomacromolecular Architecture of Eukaryotic and Prokaryotic Serine Proteases." *J. Ind. Micro.* **3** (1988):127–137.
26. Liebman, M. N., and H. Weinstein. "Heuristic Studies of Structure-Function Relationships in Enzymes—Carboxypeptidase A and Thermolysin." In *Structure and Motion: Membranes, Nucleic Acids and Proteins*, edited by E. Clementi, G. Corongiu, M. H. Sarma and R. H. Sarma. Albany:Adenine Press, 339–359.
27. Liebman, M. N. Unpublished.
28. Liebman, M. N., and T. F. Kumosinski. "Database of Protein Circular Dichroism Spectra." Unpublished Library.
29. Markley, J. Department of Biochemistry, University of Wisconsin-Madison, personal communication through National Library of Medicine, 2/89.
30. Neurath, H., and K. A. Walsh. "Role of Proteolytic Enzymes in Biological Regulation." *Proc. Natl. Acad. Sci. USA* **73** (1976):3825–3832.
31. Pethig, R. *Dielectric and Electronic Properties of Biological Materials.* New York: John Wiley and Sons, 1979.
32. PIR, Protein Identification Resource, a national resource for amino acid sequence data, maintained at NBRF, Georgetown University.

33. Prestrelski, S. J., A. H. Lipkus, and M. N. Liebman. "Analysis of Conformational Perturbation in Trypsin as Monitored by FT-IR Spectroscopy and X-Ray Crystallography." *Biophys. J.* **52** (1988):299a.

34. Prestrelski, S. J., A. L. Williams, Jr., and M. N. Liebman. "Development of a Protein Substructure Library." In preparation

35. Rose, Z., S. V. Amato, and M. N. Liebman. "Analysis of the Domain Structure of Phosphoglycerate Mutase." *Biochem. Biophys. Res. Comm.* **121** (1984):826–833.

36. Weinstein, H. , M. N. Liebman, and C. A. Venanzi. "Theoretical Principles of Drug Action: The Use of Enzymes to Model Receptor Recognition and Activity." In *Makriyannis, New Methods in Drug Research*. Barcelona: Prous, 1984, 233–246.

37. Williams, Jr., A. L., and M. N. Liebman. "Application of Dynamic Programming Approaches to Protein Structure and Function Mapping." Submitted for publication to *J. Molecular Graphics*.

K. Nakata
National Cancer Institute, Frederick Cancer Research Facility, Frederick, Maryland 21701;
present address: Department of Biomathematical Sciences, Mount Sinai School of
Medicine, One Gustave L. Levy Place, New York, New York 10029

Statistical Analysis of Nucleic Acid Sequences

We have previously developed a general method based on the statistical
technique of discriminant analysis to distinguish functional sites on nucleic
acid sequences. The attributes used for discrimination include the accuracy
of consensus sequence patterns measured by the perceptron algorithm, the
base composition and periodicity, the free energy of snRNA and mRNA
base pairing, the thermal stability map, and the Calladine-Dickerson rules
for helical twist angle, roll angle, torsion angle and propeller twist angle.
We discuss the usefulness and the further applicability of this method.

INTRODUCTION

Functional regions in nucleic acid sequences contain specific sequence patterns.
For example, the splice junctions of eukaryotic mRNA precursors contain GT and
AG pairs at the 5' and 3' termini of the intron.[1,3,25] However, GT and AG pairs
also occur at numerous other locations in the sequence, and their presence alone
is a poor predictor of splice junctions. A longer consensus example is in the *E.*

coli promoter. The *E. coli* promoter generally contains two regions homologous to "TTGACA" and "TATAAT."[11,22,27] These two regions are located about 35 and 10 base pairs upstream of the mRNA start site, respectively, and are separated by 15 to 19 bases.[29] However, a sequence containing two segments homologous to "TTGACA" and "TATAAT" with a spacing of 15 to 19 bases can also occur at numerous other locations in the genome.

A number of authors have reported statistically significant attributes of protein-coding regions and their boundaries. For example, non-random usage of degenerate codons or, as its consequence, periodic appearance of specific bases in the sequence have been used to find protein-coding regions.[7,26,28] However, such a tendency is not strong enough to reliably distinguish coding from non-coding regions. Around the promoter regions, the relative richness of A+T content is already known.[11] In our previous paper,[19] we calculated the thermal stability along the DNA sequence in terms of melting temperature and the helical twist angle, roll angle, torsion and propeller angle using Calladine-Dickerson's rules. However, each one does not dominate recognition by itself.

Another systematic approach to recognize unknown patterns, using the perceptron algorithm, was taken by Stormo et al.[30] to find beginnings of protein-coding regions in *E. coli*. The method finds by an iterative training procedure a weighting function which discriminates two sets: a true set with sequences known to have the function of interest and a false set with spurious analogues of sequences known not to have the function. The weighting function can usually distinguish between true and false sequences with 100% reliability for sequences in the original training set. However, the method is not totally reliable for sequences that are outside the training set.

Using the statistical method of discriminant analysis, we can combine any number of methods and provide the most reliable approach to distinguish true from false sequences.[13,15] If attributes are well chosen, the range of values in the true sequence will differ widely from the values in the false sequence.

METHODS

DATABASE

To distinguish the "true" sequences that surround the specified markers from the corresponding "false" sequences that surround the spurious analogues, we collect the known true and false sequences as much as possible. From GenBank, we pick up the true sequences that surround the start codon (ATG), the stop codons (TAA, TAG, TGA) and splice junction dinucleotides (GT and AG), and the false sequences that surround the spurious analogues. As for promoter sequences, we used the sequences that Hawley and McClure determined by genetic and biochemical evidence.[11] Spurious analogues were picked up from the database, so that they contain two segments homologous to TTGACA and TATAAT by more than 8 hits, with a spacing of 15 to 19.

DISCRIMINANT ANALYSIS

The basis for distinguishing true from false sequences rests on finding attributes whose distributions of values in the two sets are displaced from one another. Given an attribute vector **x** (i.e., a value for each member of the set of attributes under consideration) for a particular sequence, discriminant analysis provides a method of deciding on the population (in this case, true or false) to which the sequence is most likely to belong. Roughly speaking, this is done simply by evaluating the distribution functions for the vector **x**, in both the true and false sets, and allocating an unknown to the set whose distribution function has the larger amplitude.

Characterizing the distribution profiles against the attribute vector **x** for the known true and false sets of sequences, an unknown sequence is classified to the true set if Eq. (1) is satisfied.

$$P\left(\frac{S^+}{x}\right) > P\left(\frac{S^-}{x}\right), \tag{1}$$

where $P(S^+/x)$ and $P(S^-/x)$ are, respectively, the conditional probabilities that a sequence belongs to the true and false sets. The conditional probabilities are calculated by Bayes' theorem (2)

$$\begin{cases} P\left(\frac{S^+}{x}\right) = \frac{P(\frac{x}{S^+})P(S^+)}{[P(\frac{x}{S^+})P(S^+)+P(\frac{x}{S^-})P(S^-)]} \\ P\left(\frac{S^-}{x}\right) = \frac{P(\frac{x}{S^-})P(S^-)}{[P(\frac{x}{S^+})P(S^+)+P(\frac{x}{S^-})P(S^-)]} \end{cases}, \tag{2}$$

where $P(S^+)$ and $P(S^-)$ are the prior probabilities of true and false sets, respectively. Therefore, the condition (1) becomes Eq. (3):

$$P(S^+)P\left(\frac{x}{S^+}\right) > P(S^-)P\left(\frac{x}{S^-}\right). \tag{3}$$

If probability distributions are assumed to be normal with equal variance, Eq. (3) reduces to a linear function of x. If the distributions are normal but with different variances, Eq. (3) becomes quadratic in x. We use this quadratic discriminant function in this work.

The result of discrimination between two populations is represented by the matrix:

$$\begin{pmatrix} D_{TT} & D_{TF} \\ D_{FT} & D_{FF} \end{pmatrix}.$$

D_{TT} is the percentage of sequences predicted to be in the true category that are actual markets, and D_{FF} the percentage predicted to be in the false category that are in fact spurious analogues. The off-diagonal elements of the D matrix are related to the misclassification frequencies, a predicted marker being spurious, or a predicted spurious analogue being an actual marker. To express the overall degree of discrimination, we use a weighted average of diagonal elements D_{TT} and D_{FF}, the weight being specified by the numbers of sequences to be predicted:

$$\text{Weighted Average} = \frac{D_{TT} \times N_T + D_{FF} \times N_F}{N_T + N_F}.$$

We now describe each of the attributes that can be used, either alone or in combination with one another.

DISCRIMINANT VARIABLES

1. PERCEPTRON VALUE. The perceptron algorithm is a kind of learning algorithm for character detection.[16,24] The method was first applied by Stormo et al.[30] to analyze nucleic acid sequences. Suppose we are interested in a base sequence of length N. We encode the sequence pattern in a matrix of $4 \times N$ elements. For example, the sequence pattern of an eight-base segment, TATAATGC, is represented as follows:

$$
\begin{array}{c}
\begin{array}{cccccccc} 1 & 2 & 3 & 4 & 5 & 6 & 7 & 8 \end{array} \\
\begin{array}{c} A \\ C \\ G \\ T \end{array}
\left(
\begin{array}{cccccccc}
0 & 1 & 0 & 1 & 1 & 0 & 0 & 0 \\
0 & 0 & 0 & 0 & 0 & 0 & 0 & 1 \\
0 & 0 & 0 & 0 & 0 & 0 & 1 & 0 \\
1 & 0 & 1 & 0 & 0 & 1 & 0 & 0
\end{array}
\right).
\end{array}
$$

We denote this $4 \times N$ matrix by S and, in addition, a member from the true set by S^+ and a member from the false set by S^-. The purpose of the perceptron algorithm is to find a weighting matrix W of $N \times 4$ elements, such that Eq. (4) is satisfied.

$$
\begin{cases}
W \cdot S^+ \geq T_1 & \text{for all members of } S^+ \\
W \cdot S^- < T_2 & \text{for all members of } S^-
\end{cases}
\tag{4}
$$

where (perceptron value)

$$W \cdot S = \sum_{i=1}^{N} \sum_{j=1}^{4} W_{ij} S_{ji}. \tag{5}$$

Starting with an arbitrary initial value, the optimal W is found by an iteration. In each step Eq. (5) is calculated for all members of true and false sets, and W is modified according to:

$$
\begin{cases}
\text{if } W \cdot S^+ < T_1 & \text{then } W = W + S^+ \\
\text{if } W \cdot S^- \geq T_2 & \text{then } W = W - S^-
\end{cases}
\tag{6}
$$

The procedure is repeated until no modification is necessary; it usually converges in a finite number of steps. At this point the condition (4) is satisfied. Once the weighting matrix W is found, the perceptron value $W \cdot S$ can be used as the discriminant variable.

2. BASE COMPOSITION. The base composition, i.e., the number of bases of a particular type (A, C, G, T, A+T, G+T) divided by the total number of bases, is calculated on both sides of the boundaries of coding and noncoding or around promoter regions. The tendency of periodic appearance of specific bases at every three bases in the coding region is quantified by the following formula:

$$\text{Periodicity} = \sqrt{\left\{\sum_{i=1}^{N}(x_i - \bar{x})\cos\left(\frac{2\pi i}{3}\right)\right\}^2 + \left\{\sum_{i=1}^{N}(x_i - \bar{x})\sin\left(\frac{2\pi i}{3}\right)\right\}^2}, \quad (7)$$

where x_i (either A, C, G or T) is 1 or 0 depending on whether the base at position i is or is not of type x, and $\bar{x}$ is the average composition over a sequence of length N.

For the boundary of coding and noncoding region, we also use the procedure given by Fickett[7] to combine compositions and periodicities where each component of the latter is represented by:

$$\text{X} - \text{Position} = \frac{\max(x1, x2, x3)}{1 + \min(x1, x2, x3)}. \quad (8)$$

Here xi is the number of occurrences of a base of type x summed over positions i, $i+3$, $i+6, \ldots$ of the segment. Using our database, we recalculated the probability distribution p_i that the sequence is in a coding region for each i of the eight different parameters: four base compositions and four periodicities. An unknown sequence is classified as coding, noncoding or no opinion (don't know) depending on the value of a linear combination of these probabilities:

$$F = p_1 w_1 + \cdots + p_8 w_8, \quad (9)$$

where w_i is the percentage of the time (above 50%) that each parameter alone successfully predicted coding or non-coding function.

3. FREE ENERGY OF snRNA AND mRNA DUPLEX FORMATION. A clear pattern of complementarity to small nuclear RNAs exists in the vicinity of splice junctions.[12,14] The optimal free energy and location of snRNA base pairing to mRNA are calculated using the free energy values compiled by Salser.[23] We apply this calculation to the intron side of mRNA and snRNA of U1 type, and to the exon side of mRNA and snRNA of U2 type. The distribution profile of the optimal free energy values obtained can be used in the discriminant analysis for splice junction.

4. THERMAL STABILITY MAP. One of the statistical differences already known is the relative richness of A+T content around promoters.[11] In accordance, there have been suggestions that RNA polymerase tends to bind to low melting temperature regions,[5,31] which are likely to be easily unfolded. Using Poland's algorithm[21] with the thermodynamic parameters given by Gotoh and Tagashira,[8,9] we calculate the thermal stability map which shows the melting temperature, i.e., the temperature at 50% denaturation, against each base position.

5. CALLADINE-DICKERSON RULES. The regions where the enzyme is in contact with the DNA are strikingly homologous in space.[27] We examine this suggestion by calculating helical twist angle, roll angle, torsion angle and propeller twist angle using Calladine-Dickerson's sum function, $\sum_i (i = 1, \ldots, 4)$.[6] To enhance the derivations from regular B-DNA structure, the square of each $\sum_i$ value was calculated. The squared values of several successive angles were added and averaged to enable clustered regions of large (or small) derivations from regular B-DNA to be located.

RESULTS AND DISCUSSION

After the perceptron value converged, the perceptron algorithm achieves 100% discrimination for the training set of sequences, but it gives a considerably lower degree of discrimination for sequences outside of the training set. In contrast, the base compositition, the melting temperature and the three-dimensional features of the double helix are relatively unaffected in their power of discrimination whether sequences are within or without the training set. This observation raises a possibility that the degree of discrimination may be improved by combining the perceptron value, which is very sensitive to the choice of training set, with other variables that are more robust to its choice. This was in fact the case as shown in our previous papers. For the prediction of splice junction in human sequences,[18] the Fickett's function, free energy in the intron and exon models and the perceptron value reached 69.3%, 72.7% 60.2% and 90.9%, respectively. When the perceptron value and free energy (intron model) were used jointly, the value increased to 92.0%. For the prediction of promoter regions in *E. coli* sequences, the perceptron value reached 67.1% by itself and 75.0% when the perceptron value, base composition of G, the melting temperature and the helical twist angles were used jointly.[19] The degree of improvement is much larger in the promoter region (7.1%) than in splice junction (1.1%), because the degree of discrimination with the perceptron value is much lower from the beginning. Since the consensus is apparently weaker in promoter regions than in splicing junction, the perceptron value is more specific to the choice of a training set, lacking the ability to predict sequences outside the training set.

Around functional sites, complex factors are correlated. The most important signals reside in the surrounding nucleotide sequence[20,28] and the surrounding sequence alone is not sufficient for the prediction of the specific site. The interaction of RNA polymerase with any particular promoter will be a complex function of enzyme, promoter structure and environmental conditions.[4] For example, the DNA melting reaction does not limit the binding and the isomerization of promoter-polymerase complex if the temperature is over 25°C, but the rate of open complex formation is reduced if it's below 25°C. Buc and McClure[2] and Hawley et al.[10] suggested that the isomerization corresponds to the change in both RNA polymerase and DNA, and that the actual DNA melting step follows rapidly thereafter. More than one third of promoters compiled by Hawley and McClure[11] are close to other promoters.[17] Two (or more) promoters can be oriented in the same direction and transcribe the same gene or operon. Two RNA polymerases can bind within a common region and transcribe in opposite directions into separate genes or operations, and two RNA polymerase can oppose one another and transcribe both strands of DNA over a common interval. In these complex circumstances, the combination of discriminant variables is expected to be the best way for the statistical analysis of nucleic acid sequences. We expect that this method will become more accurate and useful as other suitable variables are identified and as the statistical database increases.

REFERENCES

1. Breathnach, R., and P. Chambon. "Organization and Expression of Eucaryotic Split Genes Coding for Proteins." *Ann. Rev. Biochem.* **50** (1981):349–383.
2. Buc, H., and W. R. McClure. "Kinetics of Open Complex Formation between Escherichia coli RNA Polymerase and the lac UV5 Promoter. Evidence for a Sequential Mechanism Involving Three Steps." *Biochemistry* **24** (1985):2712–2723.
3. Cech, T. R. "RNA Splicing: Three Themes with Variations." *Cell* **34** (1983):713–716.
4. Chamberlin, M. J., S. Rosenberg, and T. Kadesch. "Studies of the Interaction of *E. coli* RNA Polymerase Holoenzyme with Bacteriophage T7 Promoter A1; Analysis of Kinetics and Equilibria of Template Selection and Open-Promoter Complex Formation." In *Promoters: Structure and Function*, edited by R. Rodriguez and M. J. Chamberlin. New York: Praeger, 1982, 34–53.
5. Dasgupta, S., D. P. Allison, C. E. Snyder, and S. Mitra. "Base-Unpaired Regions in Supercoiled Replicative Form DNA of Coliphage M13." *J. Biol. Chem.* **252** (1977):5916–5923.
6. Dickerson, R. E. "Base Sequence and Helix Structure Variation in B and A DNA." *J. Mol. Biol.* **166** (1983):419–441.
7. Fickett, J. W. "Recognition of Protein Coding Regions in DNA." *Nucl. Acids Res.* **10** (1982):5303–5318.
8. Gotoh, O., and Y. Tagashira. "Locations of Frequently Opening Regions on Natural DNAs and Their Relation to Functional Loci." *Biopolymers* **20** (1981):1043–1058.
9. Gotoh, O. "Prediction of Melting Profiles and Local Helix Stability for Sequenced DNA." *Adv. Biophys.* **16** (1983):1–52.
10. Hawley, D. K., T. P. Malan, M. E. Mulligan, and W. R. McClure. "Intermediates on the Pathway to Open-Complex Formation." In *Promoters: Structure and Function*, edited by R. Rodriguez and M. J. Chamberlin. New York: Praeger, 1982, 54–67.
11. Hawley, D. K., and W. R. McClure. "Compilation and Analysis of Escherichia coli Promoter DNA Sequences." *Nucl. Acids Res.* **11** (1983):2237–2255.
12. Keller, E. B., and W. A. Noon. "Intron Splicing: A Conserved Internal Signal in Introns of Animal Pre-mRNAs." *Proc. Natl. Acad. Sci. USA* **81** (1984): 7417–7420.
13. Klein, P., M. Kanehisa, and C. DeLisi. "Prediction of Protein Function from Sequence Properties: Discriminant Analysis of a Database." *Biochim. Biophys. Acta* **787** (1984):221–226.
14. Lerner, M. R., J. A. Boyle, S. M. Mount, S. L. Wolin, and J. A. Steiz. "Are snRNAs Involved in Splicing?" *Nature* **283** (1980):220–224.
15. Mardia, K. V., J. T. Kent, and J. M. Bibby. "Discriminant Analysis." In *Multivariate Analysis*. London: Academic Press, 1979, 300–332.

16. Minsky, M., and S. Papert. *Perceptrons*. Cambridge: MIT Press, 1969.
17. McClure, W. R. "Mechanism and Control of Transcription Initiation in Prokaryotes," *Ann. Rev. Biochem.* **54** (1985):171–204.
18. Nakata, K., M. Kanehisa, and C. DeLisi. "Prediction of Splice Junctions in mRNA Sequences." *Nucl. Acids Res.* **13** (1985):5327–5340.
19. Nakata, K., M. Kanehisa, and J. V. Maizel. "Discriminant Analysis of Promoter Regions in *Escherichia coli* sequences." *CABIOS* **4** (1988):367–371.
20. Ohshima, Y., and Y. Gotoh. "Signals for the Selection of a Splice Site in Pre-mRNA: Computer Analysis of Splice Junction Sequences and Like Sequences." *J. Mol. Biol.* **195** (1987):247–259.
21. Poland, D. "Recursion Relation Generation of Probability Profiles for Specific-Sequence Macromolecules with Long-Range Correlations." *Biopolymers* **13** (1974):1859–1871.
22. Rosenberg, M., and D. Court. "Regulation Sequences Involved in the Promoter and Termination of RNA Transcription." *Ann. Rev. Genet.* **13** (1979): 319–353.
23. Salser, W. "Globin mRNA Sequences: Analysis of Base Pairing and Evolutionary Implications." *Cold Spring Harbor Symp. Quant. Biol.* **62** (1977):985–1002.
24. Sampson, J. R. *Adaptive Information Processing*. New York: Springer-Verlag, 1976.
25. Sharp, P. A. "Speculations on RNA Splicing." *Cell* **23** (1981):643–646.
26. Shepard, J. C. W. "Method to Determine the Reading Frame of Protein from the Purines from the Purine/Pyrimidine Genome Sequence and Its Possible Evolutionary Justification." *Proc. Natl. Acad. Sci. USA* **78** (1981):1596–1600.
27. Siebenlist, U., R. B. Simpson, and W. Gilbert. "*E. coli* RNA Polymerase Interacts Homologously with Two Different Promoters." *Cell* **20** (1980):269–281.
28. Staden, R. "Computer Methods to Locate Signals in Nucleic Acid Sequences." *Nucl. Acids Res.* **12** (1984):505–519.
29. Stefano, J. E., and J. D. Gralla. "Specific and Nonspecific Interactions at Mutant lac Promoters." In *Promoters: Structure and Function*, edited by R. Rodriguez and M. J. Chamberlin. New York: Praeger, 1982, 69–79.
30. Stormo, G. D., T. D. Schneider, L. Gold, and A. Ehrenfeucht. "Use of the 'Perceptron' Algorithm to Distinguish Translational Initiation Sites in *E. coli*." *Nucl. Acids Res.* **10** (1982):2997–3011.
31. Vollenweider, H. J., and W. Szybalski. "A Relationship between DNA Helix Stability and Recognition Sites for RNA Polymerase." *Science* **205** (1979): 508–511.

G. Christian Overton, Kimberle Koile, and Jon A. Pastor
Unisys Paoli Research Center, P.O. Box 517, Paoli, PA 19301

GeneSys: A Knowledge Management System for Molecular Biology

GeneSys is an experimental knowledge management system that explores issues in the automation of biosequence information analysis—especially with respect to structure/function relationships in gene expression. The system has a knowledge base of facts about molecular biology, an interface to molecular biology databases, and integrated analytical and reasoning tools. The first phase of the project emphasizes the formalization of knowledge about molecular structures and functions. The second phase will help automate the process of deriving new facts from the facts already existing in the databases, and thus will address the problem of analyzing the enormous body of sequence information that will be generated in very large-scale sequencing projects.

INTRODUCTION

Understanding the relationship between the structure and function of macromolecules is central to much of current research in biology. At the level of gene regulation, the problem can be seen, at least in part, as one of characterizing the interactions between defined regions of a nucleic acid sequence and regulatory proteins. Since the interactions themselves can be mediated by a range of physiological processes and are embedded in a network of concurrent interactions, the structure/function relationships governing gene regulation have proven exceedingly difficult to unravel, especially in the case of eukaryotic systems. As with virtually all problems in biology, information can be transferred—with due caution—across analogous and homologous systems both within and between species. Researchers therefore need timely access to information on the full range of systems that could potentially contribute to the understanding of their own work. However, the sheer volume of information generated in the biological sciences has led to difficulty in identifying and accessing relevant information, duplication of experiments, and serious obstacles to the identification of new biological laws and generalizations.

To cope with this "information overload" and with the sophisticated analytical methodologies required to address complex problems in biology, researchers have increasingly turned to computational support in two forms: tools for data analysis, and tools for information storage and retrieval. In the field of molecular biology, a variety of sophisticated data analysis tools have been introduced in recent years, primarily to aid in various aspects of biosequence analysis (see this volume and others[3,29,30]). In contrast, the introduction of data management tools has been slow. With the expected leap in the rate of biosequence determination, driven by large-scale sequencing projects, the problem of data management becomes critical. New approaches are needed which will not only ease storage and retrieval operations for researchers, but also will semi-automate the tasks of data analysis and the derivation of new facts from the database. Unfortunately, the computational tools necessary for these tasks are not currently available; existing tools suffer from a number of problems, among which are the inadequate representational power of existing data models, weak inferencing and query mechanisms, and poor synergy between the data storage and the data analysis components.

These difficulties have drawn increasing attention within the biology community, most notably from the National Research Council's Committee on Models for Biomedical Research[11] who proposed a "Matrix of Biological Knowledge" combining artificial intelligence, database, and information retrieval technologies to confront the problem. In this report, we outline the major components of the GeneSys project—a research vehicle for exploring the issues of knowledge representation, database management, and automated reasoning within the domain of the molecular biology of eukaryotic gene regulation. Our ultimate goal is to build a knowledge management system (KMS) composed of a knowledge base containing general information about molecular biology, an interface from the knowledge base to multiple molecular biology databases, and a suite of tools for reasoning about and analyzing the body of information. A large part of this effort is to develop systematic, formal representations for knowledge in the domain of molecular biology. Providing a formal structuring of knowledge will greatly facilitate the development of robust, general reasoning techniques in the domain. We have approached the problem of knowledge representation by building detailed models based on our current understanding of molecular structures and functions.

SYSTEM OVERVIEW

The architecture of the GeneSys system is shown in Figure 1. The major components are a knowledge base with descriptions of the structure and function of objects involved in gene regulation, a knowledge base/database interface that couples the knowledge base to one or more external databases, and facilities for querying and reasoning about the information in the system. Currently, the only database integrated with the system is a subset of GenBank that has been implemented in its

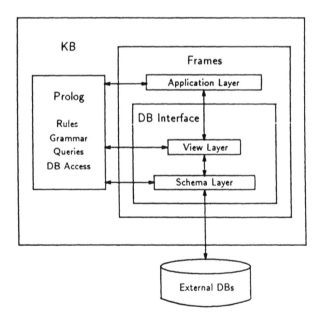

FIGURE 1   System architecture: knowledge base, knowledge base/database interface, Prolog query and reasoning facility.

proposed relational database form.[8] The major features and design criteria for each of the components are described in the following sections.

KNOWLEDGE BASE DESIGN

The issues involved in representing knowledge are complex. Concepts in physics and chemistry are relatively well defined and most knowledge in these domains is explicit; consequently, the theory, laws and generalizations in physics and chemistry usually can be expressed concisely using mathematics as a formal language and reasoned about using mathematical inference. In contrast, common-sense knowledge is not amenable to definitional representations; the relative lack of regularity, the abundance of exceptions, and the predominance of implicit knowledge render it difficult to state conditions which are necessary and sufficient for the identification of concepts. Nonetheless, representing and reasoning with common-sense knowledge is the subject of intense research and debate within the artificial intelligence community.[5] A central issue in this debate is the appropriate degree of formalism in a representation; in general, the more formal the representation, the more powerful the inferencing mechanisms that can be applied against it. Knowledge representation in biology lies somewhere between the strongly definitional physics and chemistry domains, and the largely descriptional common-sense domain; it is our goal to formalize the representation of biological knowledge to as great an extent as possible. This task is made reasonable because our focus is on the molecular biology of gene regulation, which already has a strong foundation based on biochemical principles.

Ideally, the manner in which knowledge is represented should be independent of the ways in which it is used,[14] but in practice, different organizations and structures of knowledge tend to provide support for different forms of inference. One major distinction is that between declarative and procedural representations: in declarative representations, knowledge is expressed as propositions about information, e.g., "All genes are composed of nucleic acid sequences," while in procedural representations, it is expressed as rules for managing information, e.g., "Return true for a nucleic acid sequence X if X is a gene." Declarative representations are flexible, modular, and do not necessarily imply an inference mechanism, whereas procedural representations are efficient because they have specific inference mechanisms to control search and sequencing of events. Knowledge that involves a sequence of actions is often difficult to express in a purely declarative fashion, but is natural for a procedural representation. The merits of the procedural versus declarative representation paradigms are a subject of considerable controversy.[21] Our approach is pragmatic while still striving for formalism: Structural information, such as that for a gene, can be described primarily in a declarative fashion; information about the dynamic function of biological objects is best expressed by a combination of declarative and

TABLE 1 Abbreviated Gene Frame

FRAME: Gene		
SLOT NAME	DATA TYPE	VALUE
english	String	"A region of DNA ..."
name	String	
length	Integer	
transUnit	TransUnit	
geneExpression	GeneExpression	
$\vdots$		

procedural paradigms. We are therefore using two approaches which are fundamentally declarative but have procedural support: the logic programming language Prolog and the frame language CYC. The following sections outline the knowledge representation scheme we have developed for reasoning about the structure and function of macromolecules involved in regulating gene expression.

REPRESENTING GENE STRUCTURE

Knowledge about the structure and, to a lesser extent, the function of macromolecules can be conveniently organized in a frame-based representation language. Frame languages, which have similarities to both semantic networks and object-oriented programming, have undergone considerable evolution and diversification since their introduction.[22] However, all have two characteristics in common: First, concepts in a particular domain are represented using the data-type *frame* and are described in terms of their relationships to their attributes which are represented as *slots* of the frames. Second, the frames are organized into an *inheritance hierarchy* which supports sharing of common attributes from general to more specific concepts. We illustrate these characteristics by presenting examples of concepts described in the CYC frame language. CYC[18,19] was developed specifically to address the difficult problems in representing common-sense knowledge, and therefore was designed for maximum flexibility and extensibility.

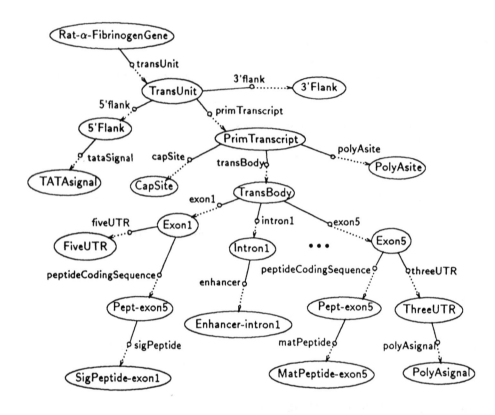

FIGURE 2   Hierarchical structure of an instance of the rat α-fibrinogen gene. The structure was deduced by analysis of the corresponding GenBank features. Ellipses represent frames. Labeled arcs are slots pointing to values of the particular gene features.

DESCRIPTION OF CONCEPTS BY ASSIGNING ATTRIBUTES. A concept central to molecular biology is that for "gene." However, "gene" has several slightly different meanings depending on the context in which it is used. Here we focus on the concept of "gene" as a region of DNA which is transcribed and whose product is either a messenger or structural RNA. The "gene" region includes the upstream and downstream sequences containing the genetic regulatory elements necessary for controlling transcription. An abbreviated **Gene** frame is shown in Table 1.[1]

The slots for **Gene** include a description (**english**), its name (**name**), its length (**length**), a representation of its structure (**transUnit**), and a representation of the

[1]By convention, CYC frame names begin with uppercase letters, slot names with lowercase letters.

processes involved in its expression (**geneExpression**). Slots in CYC are themselves frames; they therefore can have their own slots. For example, the **transUnit** slot of **Gene** is a frame that has such slots as **name** (in this case, **transUnit**) and **type** (restricting values of the slot to be of a particular type, in this case **TransUnit**, defined by another frame). A typical value may also be associated with a given slot in the context of a given frame. Finally, general procedures and local constraints can be attached to a slot to restrict slot values.

Since the attributes of a concept are also concepts represented using frames, the relationships between a concept and its attributes form a hierarchy (often called an aggregation hierarchy). The structure of the rat α-fibrinogen gene, for example, is illustrated in Figure 2. This representation makes explicit the decomposition of the structure of a gene into its substructures: the rat α-fibrinogen transcription unit is composed of a 5' flanking region, a primary transcript region, and 3' flanking region; the primary transcript region is further decomposed into a cap site, the body of the transcript, and the 3' polyA addition site; and so on. As discussed below, each frame in the hierarchy of structures for a specific gene corresponds either to a feature from the GenBank FEATURE table or to a feature derived from analysis of the FEATURE table.

DESCRIPTION OF CONCEPTS THROUGH INHERITANCE. The slots of a frame can be locally defined or acquired through *inheritance*. Slots are inherited from general frames to more specific frames via specialization links; typical values (where present) for slots are inherited from a frame to its instances. The progressive refinement of a description by specialization and modification of frames is illustrated in Figure 3. The **transUnit** slot of **Gene** has a type restriction (**TransUnit**). **Ratα-fibrinogenGene**—a specialization of **Gene**—specifies a typical value (**Ratα-TransUnit**) for that slot; instances of **Ratα-fibrinogenGene** will inherit this value (e.g., **Ratα-fibrinogenWildType1**), unless it is locally overridden (e.g., **Ratα-fibrinogenMutant1**). Because a frame can inherit slots from more than one frame, the inheritance hierarchy forms a semi-lattice.

Aside from the efficiency achieved by organizing knowledge in an inheritance hierarchy and sharing common attributes, the representation also implies a classification of knowledge: The closer two frames are in the hierarchy, the more similar are the concepts they represent. In addition, when viewed as a classification, frames higher in the hierarchy represent concepts having more general laws and rules, precisely the organization of knowledge we are seeking in structuring biological knowledge. To date, we have constructed the hierarchy entirely by hand. However, if the rules of classification can be made explicit, then the process of classification can be automated. In particular, information in the databases can be examined, and the relevant concepts can be automatically built and classified. As the size of the databases grows, automation of knowledge acquisition will become increasingly important.

CONSTRAINTS ON GENE STRUCTURE

The hierarchical structure of an individual gene, such as that for the rat α-fibrinogen gene (Figure 3), can be described using frames alone. However, since there are many different genes, an important goal of knowledge representation is to distill from the individual examples the general rules that describe permissible gene structures. In a frame language, the general rules are equivalent to constraints among the attributes of a concept.

Representing the structure of genes is very analogous to representing the structure (i.e., syntax) of sentences in natural and artificial languages, and both can be expressed conveniently as productions in a formal grammar. Brendel and Busse[7] and Searls[26] have proposed the use of formal grammars as a means of specifying the structure of genes and as a pattern-matching language to direct searches over biosequence data; our current, more modest goal is to use grammars to deduce the hierarchical structure, or *parse tree*, of a gene by examining the GenBank FEATURE

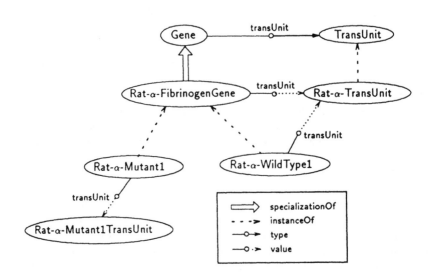

FIGURE 3   Specializations and instances of the Gene frame. **Rat_α-fibrinogen-Gene** is a specialization of the general concept Gene, **Rat_α-fibrinogenMutant-1** and **Rat_α-fibrinogenWildType1** are instances of **Rat_α-fibrinogenGene**. Graphic conventions are as described in Figure 2 with the additions shown in the legend.

```
1. transUnit(transUnit(S,E,[_5'_flank,Prim_transcript,_3'_flank]),_) —>
     feature(S,E,"locus"),
     5'_flank(_5'_flank,transUnit(S,E,_)),
     prim_transcript(Prim_transcript,transUnit(S,E,[_5'_flank])),
     3'_flank(_3'_flank,transUnit(S,E,[_5'_flank,Prim_transcript]))).
2. 5'_flank([],transUnit(Start5,prim_transcript(Start5,_-))) —> !.
   5'_flank(5'_flank(Start5,End5,[GC_signal,CAAT_signal,TATA_signal]),transUnit(Start5,_prim_transcript(SP,_-))) —>
     {End5 is SP - 1},
     gc_signal(GC_signal,5'_flank(Start5,End5,_)),
     caat_signal(CAAT_signal,5'_flank(Start5,End5,[GC_signal])),
     tata_signal(TATA_signal,5'_flank(Start5,End5,[GC_signal,CAAT_signal]))).
3. prim_transcript(prim_transcript(S,E,[Cap_site,Trans_body,PolyA_site]),transUnit(ST,ET,SFeat)) —>
     feature(S,E,"prim_transcript"),
     {contains(transUnit(ST,ET,SFeat),prim_transcript(S,E,Features)},
     cap_site(Cap_site,prim_transcript(S,E,_)),
     trans_body(Trans_body,prim_transcript(S,E,_)),
     polyA_site(PolyA_site,prim_transcript(S,E,_)).
4. trans_body(trans_body(S,E,[Exon,Intron|RestBody]),prim_transcript(SP,EP,_)) —>
     exon(Exon,prim_transcript(S,E,_)),
     intron(Intron,prim_transcript(SP,EP,[Exon])),
     trans_body(RestBody,prim_transcript(SP,EP,[Exon,Intron])).
   trans_body(trans_body(S,E,[Exon]),prim_transcript(SP,EP,_)) —>
     exon(Exon,prim_transcript(SP,EP,_)).
   trans_body(trans_body(S,E,[_5'_UTR,Pept,_3'_UTR]),prim_transcript(S,E,_)) —>
     5'_UTR(_5'_UTR,prim_transcript(SP,EP,_)),
     pept(Pept,prim_transcript(SP,EP,[_5'_UTR])),
     3'_UTR(_3'_UTR,prim_transcript(SP,EP,[_5'_UTR,Pept])).
```

FIGURE 4 Part of the grammar for describing eukaryotic protein-coding genes. The
grammar is written in Definite Clause Grammar style with a predicate term on each
line separated by ',' meaning logical AND. ',' is also used as punctuation to separate
the arguments of a term. Other symbols: '—≫' can be read as "is composed of"; '[ ]'
encloses a list and the '|' separates the head of a list from the rest of the list; '_' is the
anonymous variable; '!' is the cut symbol which provides procedural control of program
execution; '{}' specifies an escape to Prolog; lower case words and those beginning
with a number are constant symbols; and upper case words and those beginning with
an underscore are logic variables.

table. When information about the structure of a gene is transferred from the Gen-
Bank database to the knowledge base (see section on Knowledge Base/Database
interface),knowledge base/database interface the parse tree is installed as frames
representing the gene's structure as shown in Figure 2. Queries and reasoning tasks
then are addressed to the knowledge base. As the basis of our grammar, we adopt
the approach advocated by Searls and use logic grammars, written in the logic programming language Prolog, as our formalism. In the following sections, we provide
details of the grammar rules used to describe gene structure.

DESCRIPTION OF GRAMMAR. Examples of the grammar rules used to describe eukaryotic protein-coding genes are shown in Figure 4. In these rules, the surface syntax of Prolog has been changed to resemble formal grammar rules similar to those used in definite clause grammars.[10] For the most part, features in the grammar correspond exactly to the feature names and definitions proposed in the GenBank relational model.[12] The exceptions are a few additional features—transcription unit, primary transcript body, cap site—needed to fully describe the hierarchical structure of genes. The grammar rules have a one-to-one relation with the frame structures previously described: predicates are equivalent to slots and the variables in the predicate arguments are equivalent to feature types.

Ignoring for the moment the terms in braces, the grammar rules can be viewed as describing a feature on the left-hand side (LHS) of the rule in terms of its subfeatures on the right-hand side (RHS) of the rule. For example, **Rule 1** states that a transcription unit (**transUnit**) is composed of a 5' flanking (**5'_flank**) region followed by a primary transcript (**prim_transcript**) region followed by a 3' flanking (**3'_flank**) region. Similarly, **Rule 3** is an expansion of the features describing primary transcript. Terms in the grammar rules are either non-terminals such as **transUnit** (shown in **Rule 1**), which can be decomposed, or terminals, which are indivisible. Terminals can be words such as **cap_site** (shown in **Rule 3**) or the characters of the DNA alphabet—A, G, C, and T. Some of the terms which are currently terminals (e.g., **GC_signal**) may eventually be classified as non-terminals if it becomes necessary to describe their substructure in more detail.

Often an LHS can be expanded in more than one way; such disjunctions are written as individual rules with the same form of the LHS and different expansions of the RHS (e.g., **Rule 4**). Rules also can be recursive as in **Rule 4** where the LHS term, **trans_body**, appears on the RHS of the rule.

Each non-terminal predicate has two arguments: the first argument represents a frame in the parse tree and is returned by the parser, and the second argument is the parent of the current frame being examined and is used to propagate context information during the parse. The first argument of **Rule 3**, for example, is **prim_transcript(S,E,Features)** where **prim_transcript** corresponds to the type of the frame, **S** is the start position of the **prim_transcript** relative to the beginning of the sequence, **E** is the end position, and **Features** are the subfeatures of **prim_transcript**. The variables **S**, **E**, and **Features** are computed and returned by the parser.

The second argument of **Rule 3** is **transUnit(ST,ET,SFeat)** where **transUnit** is the parent node of **prim_transcript**, and **ST** and **ET** are, respectively, the start and end positions of the **transUnit**, and **SFeat** is the collection of other subfeatures of **transUnit** that have been determined to this point. The information in this argument is used to control the progress of the parse and to restrict the potential search space by enforcing restriction rules. Restrictions are non-grammar rules, and are distinguished by enclosure in curly brackets. An example of a simple restriction is the interval calculus predicate, **contains**, in **Rule 3** which checks to see if the proposed feature for **prim_transcript** is contained within **transUnit**.

RAT ALPHA-FIBRINOGEN GENE [RATFBA5E] FEATURES TABLE:

locus	1		8065	derived feature for total length of entry
TATA_signal	1723		1729	put. TATA-like sequence
prim_transcript	1739		7858	fib mRNA
precursor_RNA	1777	>	7622	fib mRNA
CDS	1835		1888	prealpha-fibrinogen, exon 1
CDS	3998		4126	prealpha-fibrinogen, exon 2
CDS	4557		4740	prealpha-fibrinogen, exon 3
CDS	5695		5840	prealpha-fibrinogen, exon 4
CDS	6483		7622	prealpha-fibrinogen, exon 5
sig_peptide	1835		1888	alpha-fibrinogen signal peptide
sig_peptide	3998		4000	alpha-fibrinogen signal peptide
mat_peptide	4001		4126	alpha-fibrinogen
mat_peptide	4557		4740	alpha-fibrinogen
mat_peptide	5695		5840	alpha-fibrinogen
mat_peptide	6483		7619	alpha-fibrinogen
intron	1889		3997	fib intron A
intron	4127		4556	fib intron B
intron	4741		5694	fib intron C
intron	5841		6482	fib intron D
enhancer	3575		3582	homology to a transcriptional core enhancer
polyA_signal	7826		7831	put. polyadenylation signal

FIGURE 5 Example of a GenBank FEATURE table. To test the grammar/parser, the GenBank and EMBL FEATURE tables for the rat α-fibrinogen gene were merged and converted to the proposed form of the GenBank FEATURE table. In addition a new feature, locus, was added to represent information about the complete sequence length. For this example, some features were ignored and features for putative enhancer and polyA signal were assumed to be confirmed.

While restrictions can be arbitrary Prolog procedures (specified by terms in braces), we are in the process of characterizing a small set that express most of the needed relations among features. Interval calculus restrictions constitute one such class. The interval calculus was originally developed as a formalism to reason about temporal intervals[1], but the same formalism can be extended easily to facilitate reasoning over intervals of biosequences by including relations to express sequence inversions, translocations, and deletion/insertions.

The GenBank parser takes as input a set of tuples for individual gene entries. Since these tuples, derived from the GenBank FEATURE table, are unordered and contain a mix of terminals and non-terminals, the parsing mechanism used is similar to a top-down, left-to-right, tabular parser.[24] Input to the parser is through the extra-grammar rule **feature**, which consults the FEATURE database and returns an instance of the required feature. The result of parsing the GenBank FEATURE table for the rat α-fibrinogen gene (shown in Figure 5) is the parse tree, represented using frames as shown in Figure 2.

The structure of eukaryotic protein coding genes can be described in a few pages of logic grammar rules. Moreover, given the existing framework we have built, it is relatively straightforward to extend the coverage to include ribosomal and small RNA genes, and most prokaryotic genes. However, additional conceptual work will be required to capture the structure of overlapping genes found in many prokaryotic and eukaryotic viruses. Another difficulty arises when trying to describe gene "fragments." Some features of a gene, such as that for the 5′ and 3′ ends of the transcribed region, are necessary for the current version of the parser to work properly.

The main goals in building the grammar and parser have been to provide: (1) a formal structuring of the information in the FEATURE table and (2) a means to determine and instantiate the structure of a particular gene in the knowledge base. The first goal amounts to an elegant and expressive extension to the frame language paradigm. Through the second goal, structures built in the knowledge base can be used for subsequent analysis of the sequence. As discussed in subsequent sections, queries can be expressed and deductions performed that depend on complex relationships among the subsequences. Additionally, running the parser is an excellent check on the consistency of the FEATURE table; if the features fail to yield a correct parse, then the entry must contain an error. The result of the parse can be used to identify the incorrect feature and to suggest possible alternatives.

REPRESENTING GENE EXPRESSION

A single unifying formalism for representing function is more difficult to achieve and will require several cycles of refinement. We start by avoiding the philosophical problem of representing function and concentrate instead on representing the processes in which biomolecules participate—in this case, the processes in the pathway for gene expression. By explicitly representing biological processes, the knowledge base is able to deduce information that is not stored explicitly in any database, e.g., information about gene expression products such as mRNA. It also can support simulation of gene expression.[17]

FRAMES FOR GENE EXPRESSION. The process of gene expression is represented by a frame called **GeneExpression**, attached to a **Gene** by means of the **geneExpression** slot. **GeneExpression** is a specialization of **Process**, as are each of the subprocesses in **GeneExpression**. As with other frames in our knowledge base, the processes are arranged in a hierarchy: the **GeneToProteinProcess**, the process that produces protein from a gene, for example, is a specialization of **Gene-Expression**, which is a specialization of **BiologicalProductionProcess**, which is a specialization of **ProductionProcess**, and so on. The following processes are represented as subprocesses of the **GeneToProteinProcess**:

1. Transcription: **TransUnit** $\longrightarrow$ **PrimaryTranscript**
2. PrimaryTranscriptProcessing: **PrimaryTranscript** $\longrightarrow$ **PrecursorRNA**
3. PrecursorRNAProcessing: **PrecursorRNA** $\longrightarrow$ **MRNA**
4. Translation: **MRNA** $\longrightarrow$ **PreProtein**

5. ProteinMaturation: **PreProtein** $\longrightarrow$ **Protein**
6. ProteinModification: **Protein** $\longrightarrow$ **Protein'**

The inputs and outputs for each process are represented as slots on that process. **Transcription**, for example, has an input slot called **transUnitIn** and an output slot called **primaryTranscriptOut** for storing a **TransUnit** and a **PrimaryTranscript**, respectively. Constraints on these slots are used to specify such relationships as "the input of a subprocess is the output of the preceding subprocess." The processing itself currently is represented as an attached procedure. **Transcription**, for example, has a **transcribe** slot in which is stored a procedure for transforming a **TransUnit**, which is a **DNASequence**, into a **PrimaryTranscript**, which is an **RNASequence**. The other objects that take part in the **Transcription** process, for example **RNAPolymerase**, are stored in a slot called **otherInputs**. When the **Transcription** process is "run" for a particular **Gene**, new objects, such as **PrimaryTranscript** and **MRNA** instances, are created. Thus, the knowledge base may contain information that is not explicitly in a biosequence database. The representation for expression of a **Gene** is shown in Figure 6. (Notice that frames representing gene structure, namely **Rat-α-TransUnit** and **Rat-α-PrimaryTranscript**, participate in the gene expression process.)

Currently, no mechanistic detail is included in the representation. Information about how the objects such as **RNAPolymerase** participate in transcription, as well as how object concentrations affect transcription, is being added. This information is being represented using multiple process levels (i.e., defining subprocesses for subprocesses and so on) and using rules such as those that state the conditions under which a particular processing step may occur.

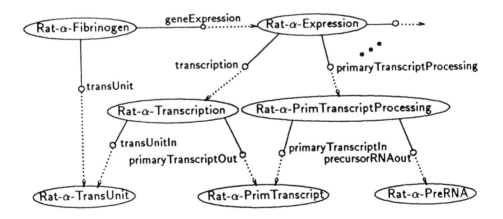

FIGURE 6 Representation for the gene expression process. Graphic conventions are as described in Figure 3.

DESCRIPTION OF RULES. Procedural information about gene expression is best represented using rules. Each rule has a premise that states conditions under which the rule holds and a conclusion that represents either a logical implication or a set of actions to be carried out. The premises and conclusions can refer to frames and slots in the knowledge base. (See Brachman et al.,[4] Vilain,[28] and Yen et al.[31] for discussions of integrating frames and rules.)

To actually "run" a particular instance of the **Transcription** process described above, the following rule, written in $P_f c$—a forward chaining extension to Prolog[13]—might be invoked:

> Rule:::concentration(Transcription,TransUnit,C_TU),
>   C_TU > 0,
>   concentration(Transcription,RNAPolymerase,C_RPol),
>   C_RPol > 0
>   ⇒
>   transcribe(Transcription,TransUnit,PrimTranscript),
>   transcriptionRate(Transcription, "> 0").

The premise of the rule states that this rule applies if the concentrations of the transcription unit (i.e., a **TransUnit**) and the enzyme RNA polymerase (i.e., **RNAPolymerase**) are greater than zero. The conclusion of the rule invokes the **transcribe** procedure, which causes an instance of a primary transcript (i.e., **PrimTranscript**) to be created and installed as the value of the **primaryTranscriptOut** slot of the **Transcription** process being run. The conclusion also sets the transcription rate for that **Transcription** process to be greater than zero. More complex rules that take into account such conditions as the presence or absence of inhibitors are described in detail in Koile and Overton.[17]

KNOWLEDGE BASE/DATABASE INTERFACE

Much of the information required by the system is contained in databases external to the knowledge base. To obtain this information, a knowledge base/database interface is needed. Ideally, an operation against the knowledge base that requires information from a database should proceed as though the information were part of the knowledge base itself. Integration of knowledge bases and databases is a nontrivial task, and is the subject of considerable study in the research community.[16] In our initial work, we have focused on integrating a single database, GenBank,[8] with the knowledge base; we will eventually extend the system to cover other databases as well.

The relational data model[9] is in many respects the most advanced available in existing commercial systems, and many current biology databases are being converted from flat files to relational form. While relational tables are conceptually simple, and the semantics of operations on them are well defined, relational tuples do not have a semantics that links them to the real world objects that they represent. A relational table is defined by a tagged set of domains. The tags themselves have no meaning to the relational database management system (DBMS), and two tables may be joined across a domain whether or not such a join is meaningful; the only restriction is that the domains are of the same data type. An additional problem with the relational model is that all hierarchical relationships are flattened—represented via joins between tables, rather than by means of hierarchical data structures. Since the knowledge base structures are intrinsically hierarchical, mapping between database relations and knowledge base structures is difficult.

INTERFACE DESIGN

The purpose of the knowledge base/database interface is to mediate and translate between these two views of the data: when an instance of a knowledge base structure that depends on the database must be built, the interface must map the knowledge base request into the appropriate series of database accesses and transformations. Design requirements for the interface included robustness, modularity, provision for multiple databases, and seamless integration of knowledge base structures derived from the database with other knowledge base structures. Both robustness and modularity are achieved by maintaining an explicit model of the database schema as part of the knowledge base: since all of the functionality of the interface is driven by this schema model, the interface will function correctly if it accurately models the underlying database. Further, by representing the semantics of the database domains within the knowledge base, the system can ensure that only meaningful operations are performed on the database. Other approaches to integrating relational databases with hierarchical data structures have been limited by failure to provide an explicit model of the interface,[2] or to integrate database-derived structures with other structures.[15]

Our first step has been to model the semantics implied by the GenBank relational model in the knowledge base by careful analysis of the proposed GenBank schema. Appropriate knowledge base concepts are then built upon the GenBank concepts by the construction of **DBViews**, which are similar to user views in a relational DBMS. This method allows the complex relations found in the knowledge base to merge smoothly with database information: the database appears as an extension of the knowledge base. We outline the structure and uses of our knowledge base/database interface below. A detailed description of its architecture and operation can be found in Pastor et al.[23]

The overall architecture of the knowledge base/database interface is illustrated in Figure 7. The knowledge base component is divided into three layers: (1) the Schema Layer, which contains descriptions of the database; (2) the View Layer, which contains descriptions of views on the database; and (3) the Application Layer, which contains application frames (e.g., **Gene**), including those wholly or partially

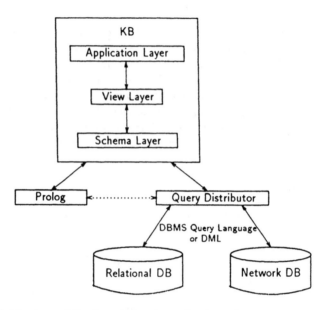

FIGURE 7 Architecture of the knowledge base/database interface.

derived from a database. The basic knowledge base structures from which the various layers are constructed are illustrated in Figure 8. A significant feature of our system is the existence of hierarchically structured views; this is in contrast to user views in the relational model, which consist of flat tuples.

Linking an application frame to the database is accomplished in three steps. First, the tables and columns in the schema must be represented in the knowledge base; semantic constraints, such as those on joins, are specified at this point. Next, application frames requiring input from the database must be created, or existing frames modified appropriately; slots are then added corresponding to database-derived attributes. Finally, the appropriate user views are created in the knowledge base by joining and projecting over tables and other views, and the application frames are linked to the views; regeneration of hierarchical structure is specified at this point via subview relationships (cf. Figure 8). As a side-effect of the creation of the views, queries corresponding to each view are generated in both Prolog and the relational query language SQL; these are stored with the view, and are used by the system when instances of database-derived frames are requested.

EXAMPLE INTERFACE

As an example of this process, we describe the mapping of a simplified notion of a gene from the database (Figures 9 and 10). The tables that we require from

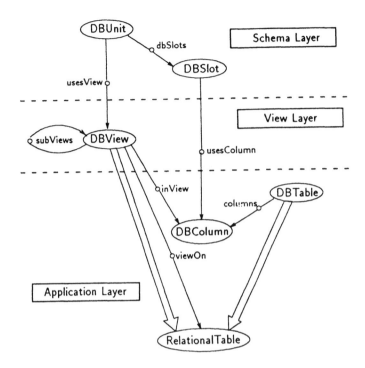

FIGURE 8 Schema layers with basic frames.

the GenBank database are GENE, GENEOCC, FEATURE, and FEATKEY (see Cinkosky et al.[8] for details on the GenBank relational schema). Frames are created for all tables and all columns, and the appropriate columns are linked with their corresponding tables. Columns over which joins are semantically valid are specified as being drawn from a common domain.

In the application knowledge base, we wish to have frames corresponding to gene occurrences, each of which has some number of features. Therefore, we create frames for **Gene** and **GeneFeature** as instances of the special frame **DBUnit**; this automatically indicates to the system that instances of **Gene** and **GeneFeature** are derived from the database. Database-derived slots are created as instances of the special frame **DBSlot**, again indicating to the system that they come from the database; these are added to **Gene**. In our simplified example, **Gene** will have slots for **geneName, geneNumber, geneOccNumber**, and **features**; note that the last of these will be filled by a set of **GeneFeature** frames, necessitating a hierarchical mapping from the database. **GeneFeature** will have slots for **name, start**, and **end**.

Finally, we create the appropriate **DBViews**. **GeneView** is a view on the GENE and GENEOCC tables, which are joined over gene number; we project

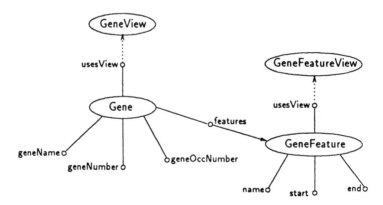

FIGURE 9 Frames for Gene and Gene Feature, with links to Views.

columns corresponding to the slots we have defined for **Gene**. **GeneFeatureView** is a view on the FEATURE and FEATKEY tables, which are joined over feature key number. While we only require three columns for **GeneFeature**, we wish to join FEATURE with the appropriate tuples of GENEOCC over gene occurrence number; we must therefore include the appropriate column in **GeneFeatureView**, so that it is accessible for an inter-view join. The application frames and slots are now linked to the corresponding views and columns, and the mapping is complete. As a result of the construction of **GeneView** and **GeneFeatureView**, the corresponding Prolog query (Figure 10) and SQL query (not shown) are generated.

INFERENCING IN THE KMS

Up to this point, we have discussed the representation of data and knowledge, and the integration of the knowledge base and database levels. The system also must retrieve and analyze data. In this section, we show how the system builds upon and extends the power of relational databases to achieve greater deductive capabilities than possible with the relational calculus. In addition to increasing the expressive power of database queries, these capabilities open the way for automated analysis of the information in the KMS.

At present, information in the KMS can be accessed either from the frame language or from Prolog. Access from the frame language is analogous to access via a user-view in a relational system, while access from Prolog, of which the relational calculus is a subset, is analogous to using a database query language. As is the case with queries to a relational database, the two methods of access offer different advantages. Queries via the frame language are simpler because they offer a pre-defined view of a segment of the KMS, but they are relatively inflexible. In contrast, queries through Prolog have the full generality of the language, but are more tedious to write and require a level of understanding that infrequent users of the system will be unwilling to obtain. Inference on the KMS falls into roughly four categories: requests for information about the KMS model, requests for objects with particular attribute values, requests for objects with constraints beyond simple attribute values (e.g., with specific relationships between constituent parts), and requests that require complex inferential reasoning.

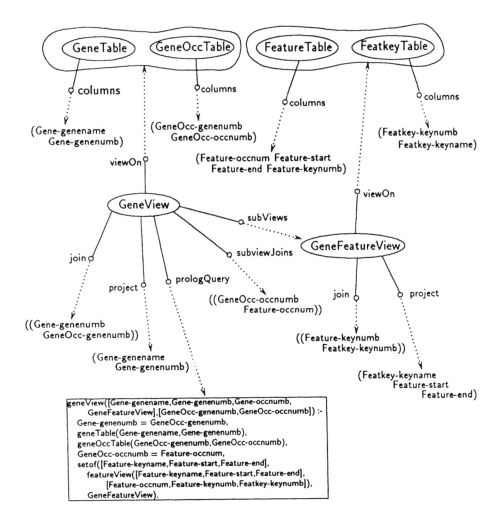

FIGURE 10 View layer. Irregular closed curves indicate multiple values for a slot.

SIMPLE QUERIES

Requests for information about the KMS model involve direct interrogation of the structure of the underlying knowledge base model, including questions about definitions of entities in the model, and about attributes of and relationships among those entities. In the frame language, such queries are equivalent to navigating through the frame hierarchy. CYC has a well-developed graphical interface to facilitate this process, the details of which can be found in Shepherd et al.[27] Programmatic queries are stated in Prolog, and can be addressed to any component of the frame system. The ability to formulate these queries obviously depends on some knowledge of the structure of frames, and of the features that can be interrogated. For example, the query "What is an intron?" would be posed as:

$$getValue("Intron", "english", Value).$$

Requests for objects with particular attribute values have direct equivalents in a relational query language. They amount to searches over the knowledge base for frames that match attribute values given in the query; this will typically necessitate a database access and is similar to a relational database query on a user view. Such queries are supported directly by the knowledge base/database interface: a request is submitted for instances of the desired type, with specifications for attributes and values, and the DBMS query is generated from a Prolog template that defines the view. The query is sent to the DBMS, and the returned values are used to instantiate the appropriate knowledge base structures (Figure 10). These queries are generated automatically by the system in response to a request for instances of a database-derived frame. For example, a request for the rat α-fibrinogen gene would be generated by a request for an instance of the frame **Gene** whose **geneName** is "rat-alpha-fibrinogen." Alternatively, these queries can be stated directly in Prolog using the view-template. For example, the Prolog query for "Find an occurrence of the rat α-fibrinogen gene" is:

$$geneView(["rat-alpha-fibrinogen," GeneNumb, GeneOccNumb, Features],_,_).$$

Queries that request objects based on complex constraints cannot be answered by simply retrieving frames that match given slot values; there are typically additional qualifying conditions that involve relationships among values in the retrieved frames. For example, to locate genes having enhancers in their first introns, it is necessary to retrieve each gene, ascertain that it has both a first intron and an enhancer, and establish that the start and end locations for the enhancer are both within those of the intron. Such queries must be formulated in Prolog, access the database through the knowledge base/database interface and return results as new instances of frames. This task is made relatively straightforward by embedding the template for the relevant view, which is already in Prolog, in a larger Prolog query. The query "Find genes with enhancers in their first introns" would be expressed as:

```
:- ḡeneView([A,B,C,Features],_,_),
      member(feature("intron",IntronStart,IntronEnd,1),Features),
      member(feature("enhancer",EnhancerStart,EnhancerEnd,_),Features),
      contains(feature("intron",IntronStart,IntronEnd,1),
            feature("enhancer",EnhancerStart,EnhancerEnd,_)),
      instantiateFrame(geneView([A,B,C,Features],_,_),Gene).
```

where the **member** predicate tests if the first argument is a member of the list specified by the second argument, **contains** is an interval calculus as previously described, and **instantiateFrame** causes a CYC frame to be created.

COMPLEX REASONING

Within the framework we have developed, complex reasoners, including complete expert systems, can be built. Inferencing systems can readily incorporate and reason with algorithmic sequence analysis tools. When algorithmic methods fail or are ambiguous, the KMS reasoner can call upon domain knowledge and facts in the database to help refine or confirm predictions based on the algorithmic methods. Derived information then can be annotated, i.e., the chain of reasoning recorded, and stored in the database. These capabilities are a step toward automated analysis of large biosequence databases.

For instance, consider the problem of identifying transcriptional regulatory elements, such as an enhancer, in a gene region. Algorithmic methods can identify putative enhancer regions, but these predictions must be confirmed experimentally. Some of the experimental evidence may reside in the database, either represented explicitly as mutants of the genes under consideration or implicitly by similarity to some homologous or analogous gene. An example of inferring the location of a regulatory element is shown in Figure 11. The rat α-fibrinogen gene has a putative enhancer element in the first intron starting at position 3575. Suppose the KMS contained information about deletion mutants of the rat α-fibrinogen gene as shown in the figure. Then the forward chaining rules (shown in the figure) would infer that indeed there was a rate-affecting region in the intron and that it was consistent with the predicted enhancer element.

Figure 3 shows a frame representation of the experimental data given in the query. **Rat-α-Mutant1** is a deletion mutant of a rat α-fibrinogen gene and is an instance of **DeletionMutant** and **Rat-α-FibrinogenGene**. Further, the transcription unit of **Rat-α-Mutant1** is a mutant of the transcription unit of **Rat-α-WildType1**. To represent this data, the two genes have different instances of **TranscriptionUnit** in their respective **transcriptionUnit** slots, and these instances are related using the slot **mutant** and its inverse **mutantOf**. **Rat-α-Mutant1** inherits unchanged attributes from **Rat-α-WildType1** by means of inheritance through the **mutant** slot. This type of inheritance is discussed in Lenat et al.[18] (The sequence missing in the mutant transcription unit is not shown in the figure, but is stored in a slot called **sequenceDifference** either on the mutant transcription unit frame, **Rat-α-M1TransUnit**, or on the mutant frame itself.) Transcription rates are represented by the **transcriptionRate** slot and may have quantitative or qualitative values. The **transcriptionRate** for **Rat-α-WildType1** is given as "X"; the **transcriptionRate** for **Rat-α-Mutant1** is given as "≪ X", a qualitative statement about its relationship to the unmodified gene's transcription rate.

Given the above representation, the rules of Figure 11, which are written in $P_f c$,[13] can be used to deduce an answer to the query. Rule **R1** is a definition of a deletion mutant and is used to deduce that **Rat-α-Mutant1** is a deletion mutant

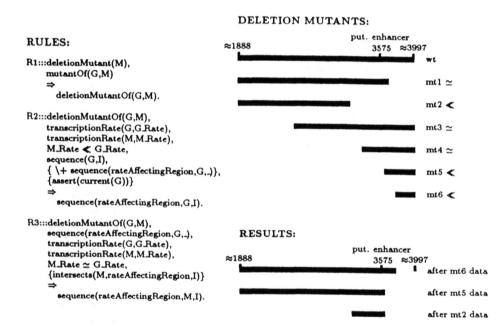

FIGURE 11 Deducing boundaries of putative enhancer region. Forward chaining rules were written in $P_f c$ to deduce the boundaries of a rate-affecting region directly from experimental data stored as mutants of a wild-type gene. To test the rules, data was entered in the knowledge base for fictional mutants having deletions in the first intron of the rat α-fibrinogen gene as shown under DELETION MUTANTS. The positions of the deletions and the corresponding rate of transcription of the mutant relative to wild type were recorded: ≈ means approximately equal to wild type and ≪ means much less than wild type. The data is similar to the results that might be expected in a deletion scanning experiment to test the transient expression of the various constructs. Results of the deduction at various stages of analysis are also shown.

of **Rat-α-WildType1**. Rule **R2** examines the mutants and notes that at least one deletion mutant causes a substantially reduced transcription rate indicating that a rate-affecting region, **RateAffectingRegion**, intersects the deleted region. Rule **R3** does all the real work by progressively incorporating information from deletions that do not greatly reduce transcription to find the most conservative bounds on **RateAffectingRegion**.

The results of the deduction are stored in the slot **possibleRateAffectingRegions** on the wild type gene, **Rat-α-WildType1**. When data for a second mutant of **Rat-α-WildType1** is entered into the knowledge base, these three rules run and again deduce possible boundaries for a **RateAffectingRegion**. In this way, as more experimental data becomes available, the boundaries of **RateAffectingRegions** in the wild-type gene can be pinpointed.[2] Because P_fc has a truth maintenance system, the deductions leading to the current state are available for examination, and if any supporting experimental data is changed or added, the system will automatically update them.

CONCLUSIONS AND FUTURE DIRECTIONS

We have described the framework for a system that can reason from experimental data to aid in the discovery of relationships between gene structure and function. It performs this reasoning by examining general knowledge represented in a knowledge base and specific information recorded in a biosequence database. The knowledge base encodes information about the linear structure of genes using frames and grammar rules, and information about the processes involved in regulating gene expression using frames and production rules. At present, the system contains only a subset of the GenBank database, but can easily be extended to include other external, pre-existing databases as long as the data can be decomposed into tuples suitable for a relational database management system or Prolog ground clauses. Various inferencing strategies can be used to reason about the contents of the system and derive new facts which can then be stored back in the database. Automated inferencing is essential for extracting the maximum possible amount of information from the large body of biosequence data that will be generated in the near future.

In the first phase of our work, we have studied the critical issues of knowledge representation as they pertain to the domain of molecular biology. A major goal of this work has been to select appropriate representation tools and to determine a pragmatic strategy for representing the information relevant to the molecular biology of gene expression. While significant problems remain unsolved in the technology of knowledge representation, AI researchers have devised a range of representation tools, from which we have selected two—logic programming and

---

[2]Our representation has been extended to include constructs that combine sequences from different genes, but these examples will not be presented here.

a frame language—for the GeneSys project, and others will be incorporated as the need arises. Having identified suitable representation tools, the critical issues at this point are what knowledge is to be represented and how is it to be structured. We originally began with the idea of representing a core of knowledge about biomolecular structures and functions, guided by the principle that the representation should be independent of how it might be used. This principle was advocated by Hayes when he proposed to represent a foundation of common-sense knowledge about the everyday physical world in order to move AI research beyond the stage of toy problems.[14] It is also central to the CYC project, whose goal is to implement a knowledge base of common-sense information that can be used by expert systems when their domain-specific knowledge fails.[19] However, after having tried to design a single, universally satisfactory representation for the domain of molecular biology, we find that in practice representation schemes must take into account intended inferencing tasks.

This observation has significantly changed our approach, and our strategy now is to select the information to be represented, motivated by the requirements of particular inference tasks. One example of a driving inferencing task was the analysis of mutants generated in a deletion scanning experiment as shown in Figure 11. While fairly simple, it led to the development of the representation of mutants shown in Figures 3 and 6 which makes explicit the relationship between the structure and function of gene regulatory regions.

It is also clear that different inferencing tasks may require different structures to represent the same information. Our work bears this out: the parser mechanism depends on the grammar rule representation for gene structure, while the knowledge-base query mechanism works best in conjunction with the frame representation. Multiple perspectives of knowledge are essential in human problem solving, and the lesson we take from our current research is that our knowledge representation system should also incorporate multiple perspectives. Thus, in addition to the above issues, we must now consider how multiple perspectives will coexist. In the case of the grammar rules and frame representation of gene structure, the two representations are complementary, and were merged by making the grammar rules constraints on the frames. Other representation problems such as those required for inferences about gene function will be more difficult to reconcile, and doing so will require a substantial effort. We now consider support for integration of multiple perspectives a central task of the GeneSys project.

The fusion of AI and database technologies described here extends the expressive power and reasoning capabilities of traditional database systems while achieving increased efficiency and persistent data storage lacking in most AI systems. We plan to enhance the system by progressively adding new inferencing tasks and techniques (e.g., pattern recognition algorithms) and the necessary supporting domain knowledge. As the system grows, the goal is to abstract and formalize the inferencing techniques and the knowledge so that combinations of techniques can be applied to new classes of problems as they arise. As a consequence, the system will become increasingly robust because it has access to a broad repertoire of inferencing techniques and the necessary supporting body of domain knowledge.

ACKNOWLEDGMENTS

We thank David Searls for expert advice and a careful reading of the manuscript. This work was supported by National Institutes of Health grant RR04026 awarded to G.C.O.

REFERENCES

1. Allen, James F. "Maintaining Knowledge about Temporal Intervals." *Communications of the ACM* **26(11)** (1983):832–843.
2. Barsalou, T., M.D. "An Object-Based Architecture for Biomedical Expert Database Systems." In *Proceedings of the Twelfth Annual Symposium on Computer Applications in Medical Care*, Washington, D.C.: IEEE Computer Society, November 1988, 572–578.
3. Bishop, Michael J., and Chris J. Rawlings, editors. *Nucleic Acid and Protein Sequence Analysis: A Practical Approach.* Oxford, England: IRL Press Limited, 1987.
4. Brachman, R. J., R. E. Fikes, and H. J. Levesque. "KRYPTON: A Functional Approach to Knowledge Representation." *IEEE Computer* **16(10)** (1983):67–73.
5. Brachman, Ronald, and Hector Levesque. *Readings in Knowledge Representation.* Los Altos, CA: Morgan Kaufmann Publishers, Inc., 1985.
6. Brachman, Ronald J., and Hector J. Levesque. "The Tractability of Subsumption in Frame-Based Description Languages." In *Proceedings of the Third National Conference on Artificial Intelligence, AAAI-84*, American Association for Artificial Intelligence. Los Altos, CA: Morgan Kauffman Publishers, Inc., 1984, 34–37.
7. Brendel, V., and H. G. Busse. "Genome Structure Described by Formal Languages." *Nucleic Acids Research* **12** (1984):2561–2568.
8. Cinkosky, Michael J., Jim Fickett, Debra Nelson, and Thomas G. Marr. "The Restructuring of GenBank." Los Alamos National Laboratory Report, Los Alamos, New Mexico, Oct 1987.
9. Codd, E. F. "A Relational Model for Large Shared Data Banks." *ACM Transactions on Database Systems* **13(6)** (1970):377–387.
10. Colmerauer, Alain, Henri Kanoui, Robert Pasero, and Phillipe Roussel. "Un Système de Communication Homme-Machine en Français." *Groupe d'Intelligence Artificiel.* Université d'Aix-Marseille II, 1973.
11. Committee on Models for Biomedical Research. *Models for Biomedical Research: A New Perspective.* Washington, D.C.: National Academy Press, 1985.

12. DDBJ, EMBL Data Library, and GenBank Staffs. "The DDBJ/EMBL/ GenBank© Feature Table: Definition Version 1.01." Los Alamos National Laboratory Report, Los Alamos, New Mexico, Sept 1987.

13. Finin, Tim, Rich Fritzson, and Dave Matuszek. "Adding Forward Chaining and Truth Maintenance to Prolog." In *Proceedings of the Fifth Annual Conference on Artificial Intelligence Applications*, Washington, D.C.: IEEE Computer Society, March 1989, 123–130.

14. Hayes, Patrick J. "The Second Naive Physics Manifesto." In *Formal Theories of the Commonsense World*, edited by J. Hobbs and R. Moore. Norwood, NJ: Ablex Publishing Corporation, 1985, 1–20.

15. IntelliCorp. "KEEconnection: A Bridge Between Databases and Knowledge Bases." 1987.

16. Larry Kerschberg, editor. *Proceedings of the Second International Conference on Expert Database Systems*, George Mason Foundation. Fairfax, VA: George Mason University, April 1988.

17. Koile, Kimberle, and G. Christian Overton. "A Qualitative Model for Gene Expression." In *Proceedings of the 1989 Summer Computer Simulation Conference*, Society for Computer Simulation, July 1989. In press.

18. Lenat, D., J. Huffman, and R.V. Guha. "CYCL: The CYC Representation Language." *MCC Technical Report Number AI-87*, MCC, 1987.

19. Lenat, D., M. Shepherd, and M. Prakash. "CYC: Using Common Sense Knowledge to Overcome Brittleness and Knowledge Acquisition Bottlenecks." *AI Magazine* (Winter 1986):65–84.

20. MacGregor, Robert M. "A Deductive Pattern Matcher." In *Proceedings of the Seventh National Conference on Artificial Intelligence, AAAI-88*, American Association for Artificial Intelligence. San Mateo, CA: Morgan Kauffman Publishers, Inc., 1988, 403–408.

21. McDermott, Drew V. "A Critique of Pure Reason." *Computational Intelligence* 3 (1988):151–160.

22. Minsky, Marvin. "A Framework for Representing Knowledge." *Artificial Intelligence Memo 306*, MIT AI Lab, 1987.

23. Pastor, Jon A., Kimberle Koile, and G. Christian Overton. "A Transparent KB-DB Interface." *LBS Technical Memo 96*, Unisys Paoli Research Center, March 1989.

24. Pereira, Fernando C. N., and Stuart M. Shieber. *Prolog and Natural-Language Analysis*. Stanford, CA: Center for the Study of Language and Information, CSLI/Stanford, 1987.

25. Schmolze, James G., and Thomas A. Lipkis. "Classification in the KL-ONE Knowledge Representation System." In *Proceedings of IJCAI-83, International Joint Conference on Artificial Intelligence, Karlsruhe, West Germany*. San Mateo, CA: Morgan Kauffman Publishers, Inc., 1983.

26. Searls, David B. "Representing Genetic Information with Formal Grammars." In *Proceedings of the Seventh National Conference on Artificial Intelligence, AAAI-88* (American Association for Artificial Intelligence). San Mateo, CA: Morgan Kauffman Publishers, Inc., 1988, 386–391.

27. Shepherd, Mary, R.V. Guha, Adolof Guzman, H. J. Hewitt, John Huffman, Doug Lenat, Chris Maeda, Claudia Porter, and David Wallace. "Updated Version: Tutorial for the Cyc Unit Editor." *MCC Technical Report Number No. ACA/AI-357-87-Q*, MCC, 1987.
28. Vilain, Marc. "The Restricted Language Architecture of a Hybrid Representation System." In *Proceedings of the Ninth International Conference on Artificial Intelligence, IJCAI-85*. Los Altos, CA: Morgan Kauffman Publishers, Inc., 1985, 547–551.
29. von Heijne, Gunnar. *Sequence Analysis in Molecular Biology: Treasure Trove or Trivial Pursuit*. San Diego, CA: Academic Press, Inc., 1987.
30. Waterman, Michael S., editor. *Mathematical Methods for DNA Sequences*. Boca Raton, FL: CRC Press, Inc., 1989.
31. Yen, John, Robert Neches, and Robert MacGregor. "Classification-Based Programming: A Deep Integration of Frames and Rules." *Technical Report ISI/RR-88-213*, USC/Information Sciences Institute, March 1988.

Karl Sirotkin and John Joseph Loehr
Theoretical Biology and Biophysics, Theoretical Division, Group T-10, Mail Stop K710, Los Alamos National Laboratory, Los Alamos, NM 87545

Simulation and Analysis of Physical Mapping

INTRODUCTION

In the introductory talk of this conference, George Bell described the various resolutions for views of a genome.[1] The present talk involves objects smaller than those that are macro-restriction mapped but larger than the bases that are sequenced. Specifically, we describe simulations of the alignment of recombinant lambdoid and cosmid clones by fingerprinting methods. The purpose of the simulation is to compare methods, as realistically as desired, while preparing for the analysis of actual physical mapping data. Furthermore, we will eventually begin to "submit" data to the Human Genome Information Resource (HGIR) to exercise its database.

A simulation has advantages over a formal mathematical analysis. Not only can a simulation be as realistic as desired (for example, by using actual sequences from GenBank[TM]) but if designed properly, when finished much of the code could be used on actual data. Furthermore, a simulation can be designed to utilize any degree of parameterization, while analyses usually must make simplifying assumptions to minimize the number of parameters. For example, the way this simulation is designed one could, by simply adding a short module, mimic rearrangements that might occur during cloning in order to discover the effect that they would have on

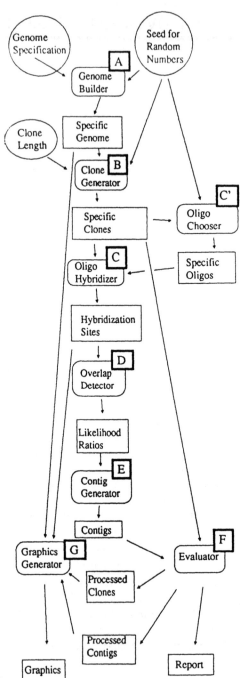

FIGURE 1 Simulation Modules. The various modules of the simulation are represented by ovals; the data files with which they communicate are represented by rectangles; and some of the user-supplied parameters are represented by circles.

the contig-generating algorithms and to learn how to recognize and deal with such rearrangements.

This talk describes the structure and announces the availability of the code for the simulation modules. We tested the method for aligning clones based upon oligonucleotide hybridizing sites, proposed by Lehrach,[5] comparing its efficacy on actual human DNA sequences from GenBank to its efficacy on random, completely uncorrelated sequences. Surprisingly, its performance was about the same on both sequences. We have also just implemented a simulation of the strategy developed at Lawrence Livermore National Laboratory that uses the occurrence of short (50 to 450) fragments. We compared these two strategies on random, structured DNA, and in general, the strategy that used the short fragments performed better, at least for some simulated conditions.

ORGANIZATION OF SIMULATION

We decided to keep the simulation as modular as possible so that we could eventually model a variety of techniques and genomes by changing a minimum of modules and supporting codes. Figure 1 represents the program modules (rectangles with rounded corners), disk files with which they communicate (rectangles), and some of the possible user-supplied parameters (ovals).

The user supplies a specification for the genomic segment from which clones are to be generated and arrayed into contigs. Currently, the pieces of the genomic segment can be: (1) random, uncorrelated bases, each of the four bases having a 25% chance of occurrence, (2) real sequences, as from the GenBank database, (3) a given sequence, mutated randomly to a given degree, (4) sequences determined by an arbitrarily long Markov process, or (5) sequences determined an n-mer at a time, with given n-mer statistics. This last "n-mer-at-a-time" process is to mimic protein coding regions, with $n = 3$. The "genome specification" (at the top of Figure 1) contains a description (for each segment of the genome) of which of the above five types is used and the mean and standard deviation of the segment's length. Additionally, relative transition frequencies to the other possible features are given. Throughout the code, a long period, linear congruence, random number generator is used. This allows reproducibility between runs; also, equivalent but different independent runs can be made by using a different starting value ("seed").

The genome building module, represented as **A** in the figure, generates a disk file sufficient to reproduce the sequences for the genomic segments. The inputs to the genome building module are the specifications for the genomic segment and the seed for the random number generator.

To generate clones from the genomic segment (module **B**), the user specifies how many clones, the mean and standard deviation for their length, the absolute limits on their minimum and maximum length, and a seed for the random number generator. Usually, for each clone three "throws" of the random number generator are required. The first gives the starting position; the second and third are used to generate the normally distributed random variable that is used in determining the length.

The selection of oligonucleotides and the simulated hybridization of the selected oligonucleotides to the genomic clones is performed by the same module, but these different tasks are separated in Figure 1 (as **C** and **C'**). There are, in general, two ways that the simulation can choose oligonucleotides. It either randomly generates and tests oligonucleotides or the oligonucleotide-choosing module uses a more complicated algorithm, which will not be further described here. If the oligonucleotides are randomly chosen, they may be either uncorrelated (with a 25% chance of each base) or generated by a Markov process of user-specified length.

The algorithm used for the simulated hybridization, searches for exact matches of the oligonucleotide sequence or its complement. It has the unusual property of being (except for the set-up described next) essentially independent of the number of oligonucleotides and of being linear in the length of the target sequence. Briefly, the algorithm notes all relevant sequence "states" or histories and how, given the next base in the sequence, to move between them. Included with each state is information about which, if any, oligonucleotide sequence is matched at the state. Another implementation of this algorithm is described in the talk given by David Torney.[7]

The same module (**C**) that simulates hybridization by string matching also notes restriction sites and predicts fragment length. To mimic experiments as actually performed, ends generated by predicted cleavage may be labelled or not. A fragment needs at least one labelled end to be detected.

No errors or intermediate levels of hybridization are yet implemented. The module that calculates the likelihood that each pair of clones overlaps (**D** in Figure. 1), uses as input the occurrence of oligonucleotide sequences in the clones or occurrence of fragments of a particular length. The method is precisely as presented[5,6] except that, when it would be impossible for the likelihood value to be above a given threshold, some values are not calculated exactly. This threshold is a user-supplied parameter (which defaults to one-third the number of clones). The filter used to avoid unnecessary calculations compares the number of oligonucleotides that actually hybridize to both clones, to the minimum number of oligonucleotides that must hybridize given the number of oligonucleotides that hybridize to one clone and not to the other. The specific numbers determined by this filter, of course, depend on the actual hybridization frequency for those oligonucleotides that hybridized to the minimum and maximum number of clones. Nesting of clones, one "within" the other, is also detected in this module. The method by which this is done is diagrammed in Figure 2. The figure shows normal overlap of two clones and the nesting of one clone in the other. The relative position of twelve oligonucleotide hybridization sites is also shown in both examples. In the case of overlap, oligonucleotides 1, 2, and 3 hybridize to the top clone and not to the bottom clone, while oligonucleotides 11 and 12 hybridize to the bottom clone but not to the top clone: the "disagreements" are thus in both "directions." In the case of nesting,

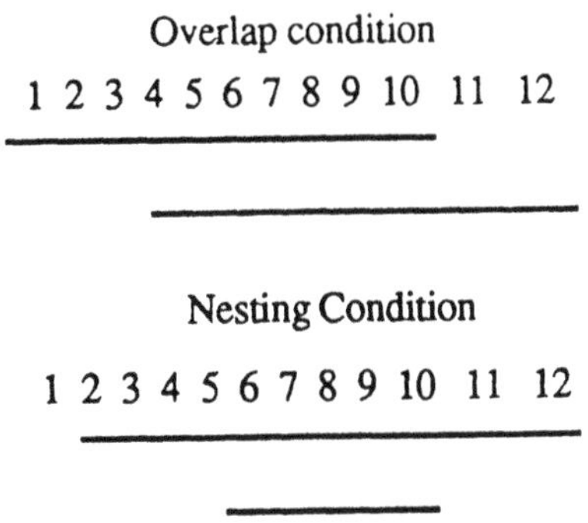

Overlap condition

1 2 3 4 5 6 7 8 9 10 11 12

Nesting Condition

1 2 3 4 5 6 7 8 9 10 11 12

FIGURE 2 Overlap and Nesting. The lines represent clone inserts and their relative positions. The numbers represent the relative position of hybridization of 12 oligonucleotides.

oligonucleotides 1 through 4, and 11 and 12 hybridize to the top clone, but there are no oligonucleotides that hybridize to the bottom clone and not the top clone: all "disagreements" are in the same "direction." The module detects nesting by the disagreements all being in the same "direction."

Although the module that generates contigs from likelihoods that a pair of clones overlaps (**E** in Figure 1) can look back to take into account clones already in the contig, a simpler "greedy" algorithm seemed to work well enough on the data sets used. This algorithm simply chose the next clone that was not nested and was not already in the contig that had the highest likelihood of overlap with the clone at the end of the contig being built. Furthermore, since our current interest is to compare overlap detection methods and since any inadequacy in contig generation would affect all methods being tested, any such inadequacy would not affect our current conclusions.

FIGURE 3 (see next page) Lehrach Method Compared Using Human Versus Random Sequences. The resulting number of contigs is plotted: the right column for 100 oligonucleotides; the left column for 70 oligonucleotides. Three results, for three equivalent runs with a different "seed" for the random number generator, are shown in each panel for each of 150, 300, 450, and 600 clones. The results for human (X) and random (O) sequences are shown. When two identical symbols would overlap, horizontal offsets are added to allow visualization of the actual number of data points. The different pairs of panels display results obtained for different conditions of contig assembly and are described in the text.

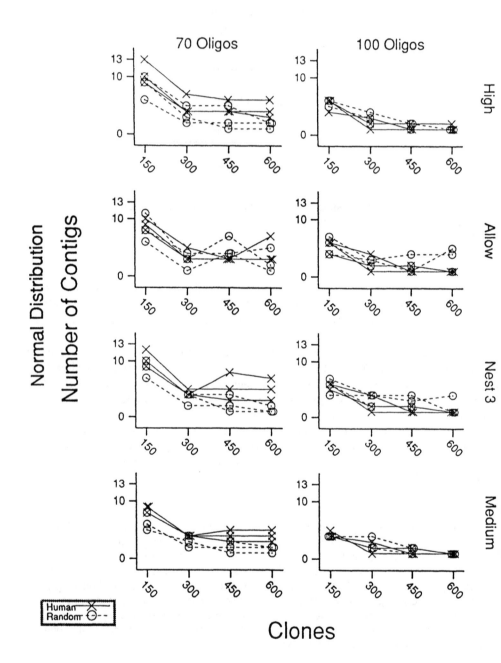

FIGURE 3

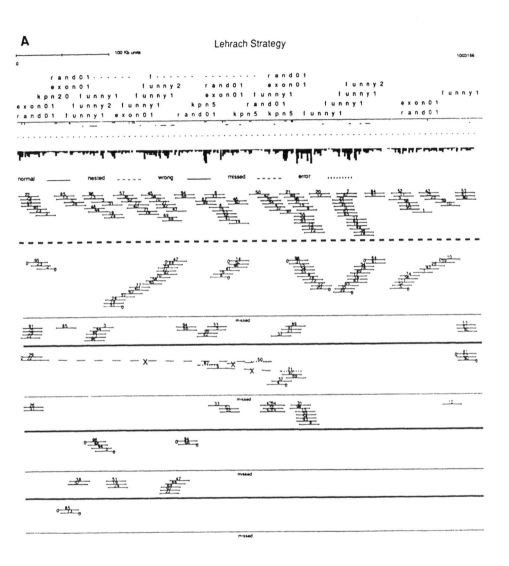

FIGURE 4 Graphical Representation of Simulation Results. (A) and (B) each portray the results of the simulation for one run using an input segment approximately one megabase long. The results for the Leharach (A) and LLNL (B) approach are directly comparable because the same clones from the same genomic segment are used in both. The top line represents the genomic segment. The names of the features used, space permitting, are above that line. The genome is generated by a Markov process in which the various "features" follow one another with prescribed transition probabilities. (continued)

B

LLNL Strategy with one 4-Hitter

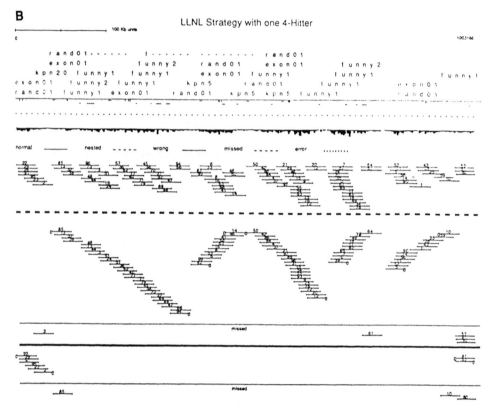

FIGURE 4   (B) (continued)

FIGURE 4 (continued) Possible features include actual sequences from GenBank, repetitive sequences—including Alu, kpn, and tandem repeats—and random nucleotides. Hypothetical genes can also be generated using codon frequencies from GenBank. An objective of this means of genome generation is to enrich the genome in possible intergene sequences relative to GenBank. The file describing the genome is reproduced in Figure 4C. Under the genomic segment, a histogram is presented showing counts of either oligonucleotide occurrence or fragments detected. The counts are on clones, so a region can be represented multiple times. This gives a measure of the relative information known at difference regions of the genomic segment. The code is then given for the iconic representations for clones under the histogram. In this figure, all but three are normal. From this region down the figure, *the horizontal positioning of the clone reflects its actual endpoints, while the relative vertical position is meaningless*, within a region of the figure. Between the histogram and a dotted line, all the clones known to the run are represented. Under the dotted line are shown contigs, the end clones of which are unambiguously identified by having a small oval on one of their ends. There are six contigs in this region in Figure 4A. Under the contigs are shown the clones missed from the *ends* of contigs. For example, clone 3 overlaps clone 51, but this overlap was missed. There is then a thick line, under which more contigs are shown, which together with their missed clones, would not have fit, without confusion, higher on the page. In the second such set of contigs and their missed clones in Figure 4A, there are features of interest. Three erroneous links occur in the same contig. For example, near the middle of the page, clone 6 was erroneously declared to overlap clone 21. This is shown by the dotted line with the "X" connecting them and clone 21 being represented by a dotted line.

C   spec random f starter
extras le 1 transitions rand01 100 -

spec random f rand01
extras le 1000 end_mean 800 end_sd 30 transitions gene01 10 funny2
5 alu10 10 alu30 20 kpn5 1 kpn20 1 intergene 80 funny1 1 -

spec markov f gene01
extras OCCURRENCE human.tri mer 3 length 200 end_mean 125
sd 5 transitions intron 90 gene01 3 exon01 7 -

spec markov f intron
extras OCCURRENCE human.di mer 2 length 2000 end_mean 500
end_sd 50 transitions intron 80 exon01 18 alu10gen 1 alu30gen 1 -

spec codon f exon01
extras OCCURRENCE human.tri mer 3 length 600 end_mean 400
end_sd 20 transitions exon01 30 intron 50 intergene 10 rand01 10 -

spec consen f alu10
extras loc mockALU len 400 mutate 10 end_mean 300 end_sd 10
transitions alu10 10 intergene 89 kpn20 0.5 kpn5 0.5 -

spec markov f intergene
extras OCCURRENCE human.tri mer 3 length 10000 end_mean
5000 end_sd 1000 transitions intergene 50 gene01 6.5 rand01 5
funny1 10 alu10 17 alu30 25 kpn5 2 kpn20 2.5 -

spec codon f funny1
extras OCCURRENCE aaaggg.tri mer 3 length 6000 end_mean
3000 end_sd 500 transitions funny2 50 intergene 50 -

spec markov f funny2
extras OCCURRENCE gatc.di mer 2 length 1000 end_mean 800
end_sd 10 transitions funny2 10 funny1 10 -

spec consen f alu30
extras loc mockALU len 400 mutate 30 end_mean 300 end_sd 10
transitions alu10 10 intergene 89 kpn20 0.5 kpn5 0.5 -

spec consen f kpn5
extras loc mockKPN len 8000 mutate 5 end_mean 6000 end_sd 100
transitions intergene 50 rand01 30 funny1 10 alu10 17 alu30 18
kpn5 1 kpn20 1 -

spec consen f kpn20
extras loc mockKPN len 8000 mutate 20 end_mean 6000 end_sd
100 transitions intergene 50 rand01 30 funny1 10 alu10 7 alu30 8
kpn5 1 kpn20 1 -

spec consen f alu10gen
extras loc mockALU mutate 10 end_mean 300 end_sd 10 transitions
intron 95 funny1 0.1 funny2 0.1 intergene 1 alu10gen 3 -

spec consen f alu30gen
extras loc mockALU mutate 30 end_mean 300 end_sd 10 transitions
intron 95 funny1 0.1 funny2 0.1 intergene 1 alu10gen 3 -

FIGURE 4 (continued) (C) The beginning of the above file is interpreted as follows (the numbers after the features named "following features" are relative transition frequencies. The actual feature is chosen depending on a throw of the random number generator. Occurrences are from human sequences from Release 55 of GenBank): *starter* one base long, always followed by *rand01*; *rand01*, random uncorrelated bases, maximum length, i.e., 1 Kb, length, otherwise normally distributed with mean 800 and standard deviation 30, followed by *gene01*—10, *funny2*—5, *alu10*—10, *alu30*—20, *kpn5*—1, *kpn20*—1, *intergene*—80, and *funny1*—1; *gene01*, chosen 3 bases at a time from actual total human occurrence, maximum length 600, otherwise mean 400, standard deviation 20, followed by *intron*—90, *gene01*—3, *exon01*—7; *intron01*, a Markov process, length 2 from actual human occurrence, maximum length 2Kb, otherwise normally distributed with a mean of 500 and a standard deviation of 50, followed by *intron*—80, *exon01*—18, *alu10gen*—1. The mockALU and mockKPN sequences bear no relationship to the actual repetitive elements but rather are random sequences 400 and 8Kb long, respectively.

The next module (**F** in Figure 1) produces various statistics by comparing the known clone positions to the overlaps declared in the preceding module. It also performs some preprocessing for the graphics module (**G** in Figure 1) that produced the graphics (Figure 4). The reader should note that the graphics module produces postscript output and/or interfaces directly with the CGS graphics software for screen display, a public domain graphics system developed and maintained at Los Alamos National Laboratory. Additionally, there is another module, not shown on Figure 1, that is used to combine information between multiple runs of the simulation and to flexibly display that information (Figures 3, 5, and 6).

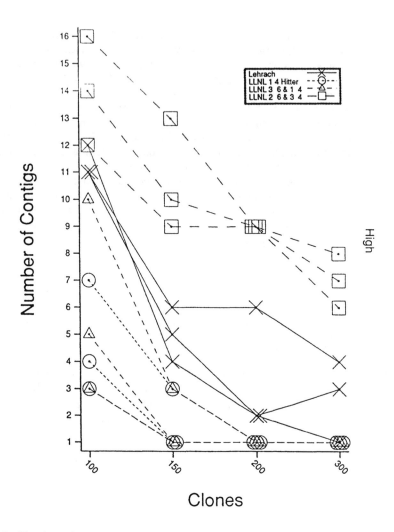

FIGURE 5 Number of Contigs from Four Simulated Protocols. The figure is laid out as in Figure 3, except all the genomic segments are as portrayed in Figure 5 and the strategies are: X—Lehrach; O—LLNL 1 restriction enzyme with a 4-base recognition site, leaving labelled ends; △—LLNL 3 restriction enzymes with 6-base recognition sites, leaving labelled ends, and 1 restriction enzyme with a 4-base recognition site, leaving unlabelled ends; and □—LLNL 2 restriction enzymes with 6-base recognition sites, leaving labelled ends, and 3 restriction enzymes with 4-base recognition sites, leaving unlabelled ends.

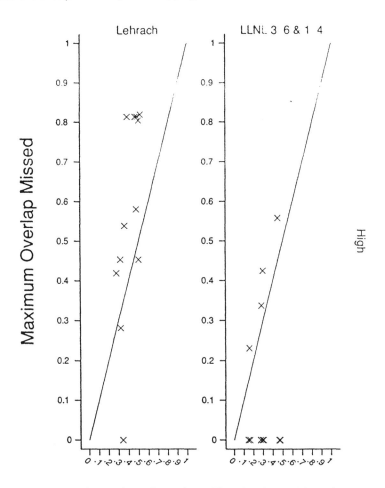

FIGURE 6　Comparison of Predicted and Expected Maximum Overlap Missed. The left panel is data from the Lehrach method; the right from LLNL 3 restriction enzymes, with 6-base recognition sites, leaving labelled ends and 1 restriction enzyme, with a 4-base recognition site, leaving unlabelled ends. The abscissa is the overlap obtained by solving for the overlap always detected in Eq. (ii′) of Lander and Waterman,[4] given the observed number of contigs. The ordinate is the observed maximum overlap missed. When theory matches experiments, points overlap the line in each panel to within 2% to 5%.

HARDWARE AND SOFTWARE

The simulation is coded in "C" and runs on SUNs using the UNIX operating system.

RESULTS

TESTING THE LEHRACH STRATEGY ON REAL VERSUS RANDOM SEQUENCES

For the genomic segment, two matched segments were used. One was composed entirely of random, uncorrelated bases; the other was composed of large human sequences and some specific repetitive sequences from GenBank release 55, separated by small, 3 Kb bases of random uncorrelated spacers. The specific human sequences used are shown in Table 1.

TABLE 1 GenBank Loci

Locus	Length	Common Name
HUMHBB	73360	beta globin cluster
HUMA1ATP	12222	anti trypsin
HUMADAG	36741	adenosine deaminase
HUMAFP	22166	alpha fetoprotein
HUMRSKPNA	3676	kpnalpha
HUMRSSAU	849	sau IIIa repeat
HUMALBGC	19002	albumin
HUMAPOBA	14070	apolipoprotein
HUMFBRG	10564	fibrinogen
HUMFIXG	38059	factor IX
HUMHBA4	12847	alpha globin cluster
HUMHLASBA	14646	hla region
HUMPARS1	11551	haptoglobin 1
HUMPARS2	10599	haptoglobin 2
HUMIFNB3	14055	interferon
HUMIL1AG	11970	interleukin
HUMNGFB	11594	nerve growth factor
HUMRSKP08	2126	Kpn repeat 08
HUMPRCA	11725	glycoprotein c
HUMTKRA	13500	thymidine kinase
HUMTPA	36594	tpa

Clone size was limited to between 15 and 20 Kb but with a mean of 17,500 and a very small standard deviation. Identical clone endpoints were chosen for the random and human sequences in order to eliminate variation in overlap as a cause of differences. (Accuracy in comparison between human and random sequences was more important than the realism of endpoints at restriction sites.) Oligonucleotides were selected from a sample of 70 clones if they had a Poisson average between 0.2 and 0.4 of "hybridizing" to a clone; oligonucleotides were used in further analysis if this average was between 0.05 and 0.55 for a given clone sample. The minimum likelihood ratio that allows a clone to be added to a contig varied and affected the number of contigs, and this is shown in different panels in Figure 3.

A matched set of results for simulation runs on both the human and random segments is presented in Figure 3. The simulation was repeated using 70 and 100 oligonucleotides, 150, 300, 450, and 600 clones. Additionally, the runs were repeated three times, with three different seeds for the random number generator: a total of 48 individual runs. Figure 3 shows the number of contigs resulting from the simulation. The left column shows the number of contigs using 70 oligonucleotides, the right column using 100 oligonucleotides. The top three pairs of panels show the results when the minimum likelihood ratio allowing a clone to be added to a contig was equal to the number of clones. They differed in that the top pair of panels ("High") shows results obtained when one "disagreement" was sufficient to remove a putatively nested clone from consideration; the next lower pair ("Allow") shows results obtained when nested clones were not removed; and the next lower pair ("Nest 3") shows results obtained when three "disagreements" were necessary to remove a putatively nested clone from consideration. The bottom pair ("Medium") is similar to the top pair, except that here the likelihood ratio allowing a clone to be added to a contig was numerically equal to one third the number of clones. Taken as a whole, the figure shows that, at least when 100 oligonucleotides were used, there was no great difference between human and random sequences. There was more fluctuation in the resulting number of contigs when using 70 oligonucleotides, and, in fact, there may have been slightly more difficulty with human sequences. The difference in contigs between the runs having the same number of clones and oligonucleotides is the result of differences in the occurrences of oligonucleotide sequences actually used.

COMPARING THE LEHRACH STRATEGY AND THE LLNL STRATEGY

For the genomic segment, one megabase segments were generated. Although three different seeds were used (and thus three different segments were used), the same segment was used for both the Lehrach strategy and the LLNL strategy. The specifications for these files are rather complex and cause the resulting genome to have structure that is at least reminiscent of real DNA in terms of having repeats the size of *Alu*'s and *kpn* repeats ("L1" repeats), stretches of degenerate tandemly repeated

DNA, and the like, and is shown in Figure 4C. Clone size and oligonucleotide selection were the same as for the previous experiment except that 100 oligonucleotides were chosen on a sample of 100 clones.

The LLNL strategy uses the occurrence of small fragments after digestion with a restriction enzyme that has a four-base recognition site. The occurrence of a fragment of a particular size and reproducible resolution to the base pair was assumed in the simulations and is treated as an event analogous to the hybridization of an oligonucleotide. In fact, occurrence of fragments from 50 to 450, inclusive, was treated after their "detection" exactly as if they were 401 oligonucleotides. The genome, clones, and resulting contigs from a matched experiment are presented in Figures 4A and 4B. In this experiment parameters are as previously described except 100 oligonucleotides chosen to be 10 bases long were used for the Lehrach strategy; for the LLNL strategy, all fragments between 50 and 450 bases produced by cleavage at GATC were noted. The clone endpoints were identical for the Lehrach and LLNL strategy and chosen so that their lengths were normally distributed with a mean of 42,500 bases and a standard deviation of 10 bases. The superior information content of the fragment occurrence used by this idealized version of the LLNL method is clearly apparent: Not only are there fewer contigs for the LLNL method, but during contig assembly, errors (erroneous links are noted with dashed lines in the figure with a large "X") were made with information obtained using the Lehrach method.

An additional experiment was performed in which 100, 150, 200, and 300 matched clones were taken from three versions of the same genome. Each version of the genome is generated using a different seed for the random number generator but the same rules (specifications). This results in completely different genomes since the order and even the relative occurrence of the features are not the same. The resulting number of contigs is displayed in Figure 5. Two additional experiments of the LLNL type were simulated. Both were simulated using one or several restriction enzymes that cleaved at six-base recognition sequences to label (or "tag") the DNA, followed by cleavage by one or several restriction enzymes with 4-base recognition sequences. Only those fragments tagged at least on one end and between 50 and 450 bases were "detected." The variability in the number of contigs when fewer clones are used is most noteworthy.

It will be useful to be able to refer to the experiments with a short name, so they are defined as follows: (a) Lehrach—with 100 oligonucleotides, (b) LLNL—with all fragments produced by one restriction enzyme cleaving at a 4-base sequence, (c) LLNL—with 3 tagging cleavages by restriction enzymes with 6-base recognition sites followed by cleavage with a single restriction enzyme with a 4-base recognition site, and (d) LLNL—with 2 tagging cleavages by restriction enzymes with 6-base recognition sites followed by cleavage with 3 restriction enzymes with a 4-base recognition site. For example, for 100 clones the variation in number of contigs was, for the various methods, (a) 11 to 12 , (b) 3 to 7, (c) 3 to 10, and (d) 12 to 16. For 300 clones it was (a) 3 to 5, (b) 1, (c) 1, and (d) 6 to 9.

Further experiments will be necessary to determine if these trends are to be expected for most cases.

Additions to the simulation will include implementing experimental uncertainty and intermediate levels of hybridization, additional algorithms for constructing contigs, and implementing the "Dupont" strategy (which is similar to the LLNL strategy except that information about the sequences at fragment ends is added).

In response to a question about the agreement between the simulations and previously published analytical predictions, Figure 6 was shown. This figure displays a comparison between actual maximum overlap missed and an inversion of the Lander and Waterman prediction.[4] The experiments were for the previous (a) and (b) type defined above. The points within 2% to 5% of the diagonal line in the figure show agreement with theory.

The method of using oligonucleotides to determine clone overlap (and simulating such use) was foreshadowed by the work of Hochman and Monahan,[2,3] who simulated the use of oligonucleotide probes for identification of viral nucleic acids.

ACKNOWLEDGMENTS

We thank Dr. David Torney for many useful discussions and Dr. Patricia Reitemeier for manuscript preparation and technical editing. This work was supported by the U.S. Department of Energy.

REFERENCES

1. Bell, George. This volume.
2. Hochberg, A. M., and J. E. Monahan. "Random DNA Probes: A Potential Method of Identifying a Virus without *a priori* Knowledge of Its Sequence." Paper presented at symposium, DNA and RNA Probes, Strategies, and Applications. Rensselaerville, NY, Sept. 6–9, 1984.
3. Hochberg, A. M. Personal communication.
4. Lander, E. S., and M. S. Waterman. "Genomic Mapping by Fingerprinting Random Clones: A Mathematical Analysis." *Genomics* **2** (1988):231–239.
5. Michiels, F., A. G. Craig, G. Zehetner, G. P. Smith, and H. Lehrach. "Molecular Approaches to Genome Analysis: A Strategy for the Construction of Ordered Overlapping Close Libraries." *CABIOS* **3** (1987):203–210.
6. Smith, George. Personal communication.
7. Torney, David. This volume.

Temple F. Smith
Dana-Farber Cancer Institute, Harvard School of Public Health, MBCRR-J815, 44 Binney Street, Boston, MA 02115

Genetic Sequence Semantic and Syntactic Patterns

INTRODUCTION

As any molecular biologist realizes, there is a nearly explosive generation of data underway in molecular genetics. Our knowledge of genetics in its most general context—gene regulation, sequence-encoded function identification, sequence evolution, genetic marker mapping and disease association, and so on—is increasing at a record rate.

It is useful to place the current data explosion in context. Biology has always been data driven. This is not a new phenomenon to biology; all one has to do is review the literature of the last century to realize that most of that literature is composed of fact, minimally synthesized or generalized by unifying principles. The famous theoretical generalizations of Charles Darwin, as expressed in *The Origin of Species*, are imbedded in long lists of observable data. Even pure "biological theorists" such as D'arcy Thompson have been forced to ideas intimately coupled with the structuring of large and often rather diverse sets of observations rather than comparison to predictive models. Thus, to some extent, biology was viewed as a science of collected and loosely organized facts until well into the twentieth century. Outside of evolution and Mendelian genetics, there were few, if any, general data reduction models or theories. There are a few unifying observations, such as "unity in diversity," as seen in the ubiquity of the fundamental metabolic cycles and the universal nature of the genetic code across diverse organisms. These, however, are not independent of the one real theory in biology, sequential historical relatedness, or evolution. Evolution is a special kind of theory. It delineates the rules by which the patterns of life will evolve, but not what those patterns will be. It is a theory not unlike some of the modern studies in chaos theory, such as the original work by Lorenz in climate patterns.

Perhaps the current data explosion has been overshadowed by our recent research experiences in biochemistry and molecular structure where there has been considerable success in applying the theoretical paradigms of the physical sciences. For many of us whose training was in the physical sciences, the new data explosion has forced us to realize that the perspective of classical physics is not applicable to much of biology. Biology is a historical science, a domain constrained by physics but not reducible to it, in the sense of global predictability.[24] This is not completely unique to biology; other large complex systems have many of the same limitations.

None of this is to suggest that the current data explosion is not real, nor perhaps different in kind. It no doubt is, and its implications are far reaching. They have appeared in the popular press, particularly in terms of the medical implications.[26] It is hard to realize, for example, that it has been less than six years since the AIDS agent was identified as a retrovirus, HIV-1, yet we now know its complete genetic sequence and that of a couple dozen variants, as well as the detailed mode of action of many of its gene products and most of its clinical life cycle.[19] This rate of data generation is nearly typical of modern biology, and the future promises only more. With the planned mapping (and sequencing) of the entire human, *E. coli*, and nematode genomes, we are in danger of being overwhelmed by raw data. We need to remember that we humans have dealt effectively with large data sets. Table 1 lists a number of data sets of interest, many of which have been organized by subject, indexing, and function (entertainment, spatial, commercial, etc.). Clearly, there is a pressing need to organize this genome data, develop methods to search it efficiently, and cross-correlate it with the wealth of related biological and medical data.

Again, however, our perspective is important. Consider the analogy between nature's language of molecular genetics and human language, a simple analogy often used in beginning biology classes. Such analogies are obvious at some level; for example, consider DNA as a language of protein sequences punctuated by promoters and poly-A attachment signals, and composed of three-letter codon words written in the four-letter DNA alphabet. Although this analogy to simple written language clearly does not seem to have the richness of even the simplest of human languages, our knowledge of genetic sequences has revealed a very complex and subtle structure to the molecular language which is at least as complex as our own mode of expression. This is not surprising. First of all, the genetic language must describe all of the intermediate metabolism and its regulation—and, as any student of biochemistry knows, most meaningful subsets of the biochemical pathways are as complex as any English sentence ever diagrammed in freshman composition!

TABLE 1 Approximate Sizes of Data Sets of Human Interest[1]

Information Source	Number of Characters	Bits
Einstein's Special Relativity[2] *Paper Ann. der Physik* **17** (1905)	0.048×10^6	0.24×10^6
Plato's Dialogues	1.4×10^6	5.9×10^6
Chemical Handbook and Catalog (Aldrich Inc.)	1.7×10^6	8.7×10^6
E. coli genome	4.2×10^6	8.4×10^6
H. G. Well's Outline of History	2.8×10^6	11.8×10^6
Shakespeare's Complete Works[3]	6.1×10^6	2.7×10^7
List of all "named" species	12.0×10^6	4.8×10^7
GenBank (nucleotides) 1988 release 58	24.7×10^6	4.9×10^7
Oxford English Dictionary	18.0×0^6	8.1×10^7
Four years of Biology major's undergraduate text books	20.0×10^6	1.0×10^8
My PC disk drive	21.0×10^6	0.17×10^9
The Talmud	1.9×10^8	$.86 \times 10^9$
A typical professor's office	0.21×10^9	1.0×10^9
Human genome[4]	3.0×10^9	6.0×10^9
U.S. personal income tax data filed in 1987	10.0×10^9	39.0×10^9
A University Library (10,000 volumes)	800.0×10^9	3.9×10^{12}
The U.S. Library of Congress	$> 10.0 \times 10^{12}$	$> 52.0 \times 10^{12}$

[1] As individuals we seem to be able to deal with very large data sets: on the order of 10^8 or so (that is, most of us can retrieve some fact, equation or quote incompletely remembered by a non-random search of our personal libraries), and corporately we handle a thousand times that routinely.

[2] Manuscript copy sold in 1944 for 6.00×10^6 (or 120 "1988" dollars per bit!).

[3] English text values used between 4.5 and 5.5 bits per character which includes a character frequency weighting of $f\{\log f\}$.

[4] The 1988 NRC/NAS report on Mapping and Sequencing the Human Genome estimated the cost at about 3.00×10^9 (or about 0.5 "1988" dollars per bit!).

Secondly, molecular genetics represents a living language with a long evolutionary history of many interacting dialects, again not unlike human languages with their cultural and historical intertwinings.

The words in this language appear to consist not just of the three-letter codons, but of functional domains such as those delineated by the exon/intron mosaic nature of eukaryotic coding regions.[12,13] These domain words include all of the protein functional domains, as well as the various regulatory elements, repeats, spacers, recombination hot spots, replication origins, and so forth. The latter group of regulatory elements covers an expanse as great as the former: included are promoters, ribosomal attachment sites, and many other elements either not yet identified or whose function remains obscure. One current view is that this mosaic structure, first observed in the exon/intron structure, allows for the shuffling of protein domains while minimizing the chance of scrambling the individual function domains (exons) or words. In this view, the introns and, more generally, all non-explicitly encoding or regulatory domains are shuffling spacers—from the evolutionary point of view, providing a statistical advantage similar to sex.

As in human language, the meanings, connotations, and the spelling of words evolve slowly compared to their allowed syntactic usage. For example, once the exons are duplicated and/or rearranged, they are under new evolutionary pressures and will evolve independently of their previous genetic sentence context. Note, for example, the so-called homeobox containing genes[14] that perform related regulatory functions in man and Drosophila, but do so within very different embryonic development schemes.

The complex syntactic structure of this language includes relational units analogous to the parts of speech. There are, for example, verb-like words: the three different eukaryotic RNA polymerase promoters, the DNA replication origins, and perhaps we should include the various protein active sites. There are adverb-like words, including enhancers and protein leader signal peptides. The various promoter effectors, such as the SP1 protein[7] or the large family of DNA zinc finger-binding proteins,[10] can form the equivalent of large adverbial phrases. Consider the possibilities given a set of a dozen or so short recognition signals, any three or four of which—if contiguously bound by their respective regulatory protein—form a promoter activator. This would give thousands of combinations, each generating various degrees of activation. These in turn allow large overlapping suites of genes to be co-regulated in concert, a requirement for cellular differentiation and the complex embryonic sequences. Given the potential for the various regulatory proteins to be self- or co-regulated through positive and negative feedback, this language can encode complex conditional conjunctive logic. We currently see only the glimmerings of such structures, for example, in the cell-cell interactions among the various "helper," "effector," and "suppressor" T-cell functions in the immune system. However, the means of encoding the wondrous spectrum of spatial and temporal patterns found in modern living organisms can now be clearly envisioned.

In order to facilitate the investigations of the semantic and syntactic structure of this genetic language, two major efforts are currently receiving strong support: (1) the continued generation of more sequence and structural data and (2) its organization into modern databases. There are three major molecular databases: GenBank, the nucleic acid sequence database[3]; the NBRF Protein Information Resource (PIR), containing amino acid sequences[11]; and the Brookhaven National Laboratory protein x-ray structure database. There are also a number of related databases, including, for example, the Yale "Human Gene Mapping Library" and the database of "Sequence of Proteins of Immunological Interest."

SEQUENCE COMPARISONS

The Rosetta Stone of modern biology appears to be sequence comparative analysis. The most common request made of either GenBank and/or the PIR, for example, is for a list of all sequences similar to a newly sequenced query along with some measure of the relative statistical significance of the similarities. The identification of such sequences by name or even simple description is, however, just the start. One normally wants to know if any of these potentially similar sequences share a biochemical function or pathway, if they occur in similar differentiated tissues, if any of their three-dimensional structures are known, if their genome map positions are known, if they share common regulatory signals, if they are implicated in any clinical disorders, and finally what recent research literature cites these sequences. Some of the other challenges are to develop data and terminology cross-references (data field and nomenclature thesaurus or synonym dictionaries). Another is to develop relational database algebras that will operate on numerous independently structured databases via such cross-referencing and thesaurus intermediates. The aim is to allow initial searches for particular relationships (a sequence similarity or implied homology) between a new DNA sequence and those in the current databases, which will then result in the identification of all other intersections (in the set theory sense) with associated data. Such a tool would result in the potential for the identification of the new, often unanticipated, emergent relationships so essential to science.

General methods are not yet available to uniquely identify all elements in the primary sequence or encoded structures associated with a particular biochemical function. What is needed are methods of extracting from a set of sequences all encoding for some common function, a common pattern if there is one. This is, of course, one of the major challenges.[5]

STRUCTURAL MOTIFS

We (Lathrop et al.,[15] Smith and Smith,[23] and Webster et al.,[28]) have been working on developing tools that can help in identifying functional pattern for common structural motifs. The first of these tools exploits the inherent evolutionary structure of the databases, while the second attempts to induce structure on the databases through the generalization of functional correlate patterns. Families of related protein sequences in the NBRF/PIR database were identified by running the entire database against itself using DASHER, MBCRR's high-speed similarity search program. Homologous families were defined as clusters of sequences all having a pairwise DASHER Chi-squared score of 6.0 or greater. (The Chi-squared measure here is a measure of deviation in the maximum aligned similarity from all other align similarities between the given sequence pair.) A 50% linkage rule was then used to generate 1067 clusters of 2 sequences or more, encompassing 4950 of the 7396 sequences in NBRF (v.16). These clusters include, for example, beta hemoglobin, alpha-hemoglobin, Ig kappa chain, cytochrome c, actin, insulin, and tubulin sequence families. For each of these clusters, a single "consensus pattern" has been constructed by (1) optimally aligning the two most similar sequences in the cluster, (2) replacing the align sequences with a common consensus-like sequence in the cluster hierarchy, and then (3) repeating steps 1–2 for the next most similar pair of sequences (either actual or consensus-like sequences) until a "root" consensus pattern for the entire cluster is generated.

From each pair of aligned sequences the single consensus-like pattern is constructed by substituting a single character for each pair of aligned residues. If identical residues are aligned, that residue is placed in the consensus sequence. Otherwise, a pre-defined amino acid class hierarchy is used: (((([DE],[KRH,[NQ], [ST]) [DEKRHNQST]), ((([ILV],[FWY],C,P,M) [ILVFWYCPM]),([AG])),X), and the minimally inclusive class is assigned to the aligned position. Gaps in the alignment are replaced by gap-characters in the consensus sequence, and the associated gap-penalty is reduced within these regions upon subsequent alignment. Since the minimum inclusive amino acid class is represented at each position of an optimal (maximum similarity) alignment, these "consensus" sequences are properly termed "minimal amino acid class covering" patterns. This is to be distinguished from the standard "majority-rule" consensus. An example of the homology family and the associated covering pattern is shown in Figure 1.

In addition to their diagnostic use in similarity searches, sub-patterns representing conserved amino acid sites/domains derived from these amino acid class coverings will be useful in gene cloning and protein engineering studies.[21] Reverse translation of stretches of conserved residues can be used to design degenerate pools of hybridization probes. Such probes could then be used to clone the corresponding gene in other organisms. In addition, since conserved sites/regions most likely represent functionally important residues and/or domains, such patterns will aid in the identification of useful sites for *in vitro* mutagenesis studies relating structure and function.

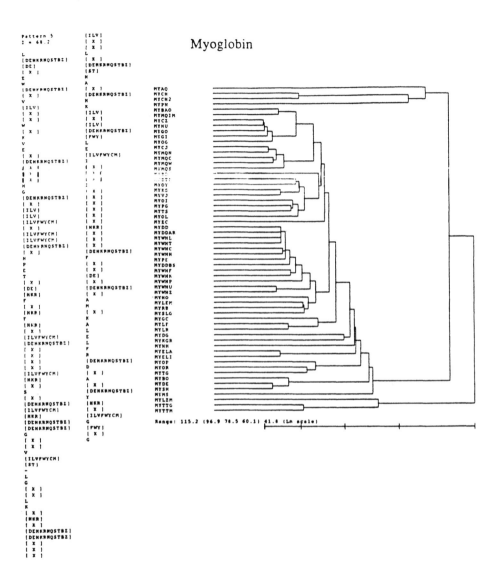

FIGURE 1   An example of the myoglobin homology family and the associated covering pattern (from the analysis of R. Smith[23]).

AN EXAMPLE: THE MONONUCLEOTIDE BINDING FOLD

The 'mononucleotide binding fold' (MNBF) is one of the most common and best studied of such structural/functional protein motifs. Its existence in the aminoacyl-tRNA synthetases, as well as in the NAD and FAD-binding group, suggested that a computer-assisted search for those combinations of primary and predicted secondary structural characteristics that represent an MNBF[22] might lead to the detection of previously unrecognized nucleotide binding domains, where simple searches for common primary sequences, alone, had failed. Operating on this hypothesis, we have recently developed a new protein pattern-matching system, ARIADNE,[15] and used it to locate a Rossman-type MNBF in proteins of disparate function whose X-ray structures have not yet been solved.[28]

The system employs sets of pattern descriptors composed of combinations of common primary sequence and predicted secondary structure known to constitute a Rossman-type MNBF in members of a particular protein class where at least some three-dimensional information is known. A different descriptor was developed for each class of nucleotide-binding protein studied. ARIADNE was used to provide the optimal alignment of the descriptors against sets of protein primary sequences annotated with secondary structure and hydropathy profiles. The procedure allows for both conflicting secondary structure annotations and the uncertainties in all such prediction methods. When this system was applied to a search for a consensus MNBF in a set of 12 aminoacyl tRNA synthetases, it predicted the existence and location of such a structure in 67% of these enzymes; moreover, the criteria applied to the construction of the relevant pattern descriptors were sufficiently stringent that only four false positives were selected from a set of 54 control proteins of disparate function, studied in parallel.

We have constructed and tested descriptors for structural motifs used for the binding of different nucleotides. Nucleotide-binding X-ray structures share the following fold: a center beta-strand followed by a turn and an alpha-helix which lies parallel to the plane of the beta-strand (reviewed by Bradley et al.[6]). The invariant amino acids and the structure of the turn varies depending on the class of bound

TABLE 2 Classes of Bound Nucleotides and Their Descriptors[1]

Nucleotide	Descriptor
NAD/FAD	Strand-A GXXX G Helix-B $X_{-7,9}$ Strand-B E/D
ATP/ADP	Strand-A GXXX GX Helix-B $X_{0,11}$ Strand-B
ATP/AMP-AA	Strand-A $X_{0,4}$ turn X_3 G[H,N,Q] Helix-B $X_{0,11}$ Strand-B

[1] Higher-order elements are enclosed in boxes. "Spacer" elements are X_{ij} where i and j are the allowed range of amino acids of any type (X). A negative value is the length of an allowed overlap of predicted secondary structure boundaries. Each of the primary sequence elements has essential structural or functional roles.[6,28]

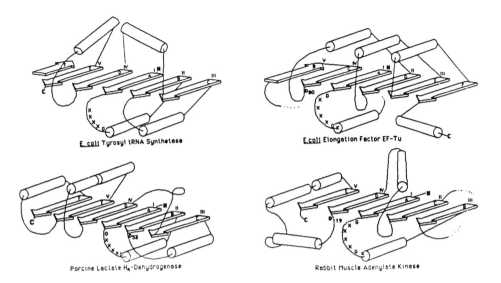

FIGURE 2   Schematic diagrams of the known nucleotide-binding sites of proteins from four different nucleotide binding classes. The diagrammatic method of Zelwer et al.[29] was used for the graphic presentation of selected secondary structural components of proteins whose three-dimensional structures are known. Arrows represent B-strands that are pointed in the C-terminal direction, and cylinders represent α-helices. Bold letters N and C indicate the amino and the carboxyl ends of the partial sequences presented. B-strands are labeled alphabetically to indicate their relative position in the primary sequence. Five B-strands (A–E) were assigned a Roman numeral according to their relative positions in the observed B-sheet. Of these 5 B-strands, A was found to occupy the central position (I) in each case and was the first element of a Rossman-MNBF (b-strand—GXXXG—a-helix—(0–11aa)—B-strand). Figure adapted from Figure 2 in Ref. 6.

nucleotide (see Figure 2). These classes and their descriptors are shown in Table 2. A search for matches to these descriptors within several functional groups was executed by the pattern-directed inference system, ARIADNE. As discussed below, each of these showed high sensitivity and specificity. Also, a highly specific descriptor based on the *Escherichia coli* Met-tRNA synthetase binding fold for *E. coli* Ile-tRNA synthetase. A portion of this model has been experimentally validated in a collaborative study.

To validate the method, we constructed a single descriptor that specifically locates common structural motifs within dissimilar primary sequences, using only information inferred from the primary sequence. The above NAD/FAD descriptor was constructed from the alignment of common structures shared by eight dinucleotide-binding oxidoreductases.[4] This was tested against the primary sequences annotated with predicted secondary structures of the same eight proteins. Its specificity is 75%

(6 proteins with correct matches/8 proteins). 86% of the matches in the oxidoreductase set were true positives (6 correct matches/7 matches). One of seven matches did not locate a nucleotide-binding site. This experiment was also a test case for the indirect method of estimating the false positive background, which compares with the percent of matched control proteins (52 control proteins with no matches/54 control proteins). It validates the intuition that when the specificity for the functionally related proteins (75%) and the sensitivity based on the control set (96%) are both high, the majority of the matches located in the functional set (86%) are in fact true positives.

The following version of the above NAD/FAD descriptor located a potential nucleotide-binding site within the human mitrochondrial URF6 gene product.

Strand-A   GXGXX   G Helix-B $X_{-7,9}$ Strand-B

Our analysis corroborates the recent proposal of Chomy et al.,[9] in which they identified the URF6 product as being a subunit of NADH dehydrogenase, using antibodies against predicted URF6 peptides. Our analyses strongly support this since the URF6 product was the only proposed NADH dehydrogenase subunit in which we are able to identify a probable nucleotide-binding domain. The seven URF subunit primary sequences were annotated with secondary structure predictions using a Chou and Fasman[20] prediction scheme. A match to the descriptor is found only in the URF6 submit product. The specificity of this descriptor is 50% (5 correct matches/10 proteins) for the known nucleotide-binding sites in the set of oxidoreductases. The sensitivity is 94% for the control proteins (49 control proteins with no matches/50 control proteins).

Much work must be carried out before one will be able to decide whether tools such as these are of general utility. The generation of functional descriptor libraries is one of the many computer-assisted studies that will be carried out over the next few years, and such information will have to be incorporated into our ever-expanding databanks.

REFERENCES

1. Anderson, J. E., M. Ptashne, and S. C. Harrison. "Structure of the Repressor-Operator Complex of Bacteriophage 434." *Nature* **326** (1987):846–852.
2. Berg, J. M. "Potential Metal-Binding Domains in Nucleic Acid BindingProteins." *Science* **232** (1986):485–486.
3. Bilofsky, H. S., C. Burks, J. W. Fickett, W. B. Goad, F. I. Lewitter, W. P. Rindone, C. D. Swindell, and C. S. Tung. "The GenBank Genetic Sequence Data Bank." *Nucl. Acids Res.* **14** (1986):1–4.
4. Birktoft, J. J., and L. J. Banaszak. "Structure-Function Relationships among Nicotinamide-Adenine Dinucleotide Dependent Oxidoreductases." In *Peptide and Protein Reviews*, edited by M. T. W. Hearn. Vol. 4. New York: Marcel Dekker, 1984.
5. Blundell, T. L., B. L. Sibanda, M. J. E. Sternberg, and J. M. Thornton. "Knowledge-Based Prediction of Protein Structures and the Design of Novel Molecules." *Nature* **326** (1986):347–352.
6. Bradley, M. K., T. F. Smith, R. H. Lathrop, D. M. Livingston, and T. A. Webster. "Consensus Topography in the ATP Binding Site of the Simian Virus 40 and Polyoma Virus Large Tumor Antigens." *Proc. Natl. Acad. Sci. USA* **84** (1987):4026–4030.
7. Briggs, M. R., J. T. Kadonaga, S. P. Bell, and R. Tjian. "Purification and Biochemical Characterization of the Promoter-Specific Transcription Factor SP1." *Science* **234** (1986):47–54.
8. Charnay, P., R. Treisman, P. Mellon, M. Chao, R. Axel, and T. Maniatis. "Differences in Human α and β Globin Gene Expression in Mouse Erythroleukemia Cells: The Role of Intragenic Sequences." *Cell* **38** (1984):251–263.
9. Chomy, A., M. W. Cleeter, C. I. Ragan, M. Riley, R. F. Doolittle, and G. Attardi. "URF6, Last Unidentified Reading Frame of Human mtDNA, Codes for an NADH Dehydrogenase Subunit." *Science* **234** (1986):614–618.
10. Evans, R. M., and S. M. Hollenberg. "Zinc Fingers: Gilt by Association." *Cell* **52** (1988):1–3.
11. George, D. G., W. C. Barker, and L. T. Hunt. "The Protein Identification Resource." *Nucl. Acids Res.* **14** (1986):11–16.
12. Gilbert, W. "Gene-in-Pieces Revisited." *Science* **228** (1983):823–824.
13. Go, M. "Modular Structural Units, Exons, and Function in Chicken Lysozyme." *Proc. Nat. Acad. Sci. USA* **80** (1983):1964–1968.
14. Hart, C. P., A. Fainsod, and F. H. Ruddle. "Sequence Analysis of the Murine Hox-2.2, -2.3 and -2.4 Homeo Boxes: Evolutionary and Structural Comparisons." *Genomics* **1** (1987):182–195.
15. Lathrop, R. H., T. A. Webster, and T. F. Smith. "ARIADNE: A Flexible Framework for Protein Structure Recognition." *Communications of the ACM* **30** (1987):909–921.

16. Lichtsteiner, S., J. Wuarin, and U. Schibler. "The Interplay of DNA-Binding Proteins on the Promoter of the Mouse Albumin Gene." *Cell* **51** (1987):963–973.
17. Marx, J. L. "A Parent's Sex May Affect Gene Expression." *Science* **239** (1988):352–353.
18. McClarin, J. A., C. A. Frederick, B. Wang, P. Greene, H. W. Boyer, J. Grable, and J. M. Rosenberg. "Structure of the DNA-Eco RI Endonuclease Recognition Complex at 3 A Resolution." *Science* **234** (1986):1526–1541.
19. Quinn, T. C., J. M. Mann, J. W. Curran, and P. Piot. "AIDS in Africa: An Epidemiologic Paradigm." *Science* **234** (1986):955–963.
20. Ralph, W. W., T. Webster and T. R. Smith. "A Modified Chou and Fasman Protech Structure Algorithm." *CABIOS* **3(3)** (1987):211–216.
21. Reichardt, J. K. V., and P. Berg. "Conservation of Short Patches of Amino Acid Sequence Amongst Proteins with a Common Function but Evolutionarily Distinct Origins: Implications for Cloning Genes and for Structure-Function Analysis." *NAR* **16** (1988):9017.
22. Rossman, M. G., D. Moras, and K. W. Olsen. "Chemical and Biological Evolution of a Nucleotide-Binding Protein." *Nature* **250** (1974):194–199.
23. Smith, R. F., and T. F. Smith. "The Generation of Diagnostic Protein Family Patterns." Submitted to PNAS.
24. Smith, T. F., and H. J. Morowitz. "Between Physics and History." *J. Mol. Biol.* **18** (1982):265–282.
25. Smith, T. F., W. W. Ralph, M. Goodman, and J. Czelusniak. "Codon Usage in the Vertebrate Hemoglobins and Its Implications." *Mol. Biol. & Evol.* **2(5)** (1985):390–398.
26. Time Magazine. "Gene of the Week." *Time Magazine* March 30 (1987):62.
27. Waterman, M. S. "General Methods of Sequence Comparison." *Bull. Math. Biol.* **46** (1984):473–500.
28. Webster, T. A., R. H. Lathrop, and T. F. Smith. "Prediction for a Common Structural Domain in Aminoacyl-tRNA Synthetases through Use of a New Protein Pattern-Directed Inference System." *Biochemistry* **26** (1987):6950–6957.
29. Zelwer, C., J. L. Risler, and S. Brunie. "Crystal Structure of *E. coli* Methionyl-tRNA Synthetase at 2.5 A Resolution." *J. Mol. Biol.* **155** (1982):63–81.

Index